Output Measurements for Medical Ultrasound

ROY C. PRESTON (ED.)

Output Measurements for Medical Ultrasound

With 75 Figures

Springer-Verlag
London Berlin Heidelberg New York
Paris Tokyo Hong Kong
Barcelona Budapest

Roy C. Preston, BSc, PhD, C Phys, F Inst P
Division of Radiation Science and Acoustics, National Physical Laboratory, Teddington, Middlesex TW11 0LW, UK

ISBN-13:978-1-4471-1885-5 e-ISBN-13:978-1-4471-1883-1
DOI: 10.1007/978-1-4471-1883-1

British Library Cataloguing in Publication Data
Output measurements for medical ultrasound.
1. Ultrasonography
I. Preston, Roy C.
616.07543
ISBN-13:978-1-4471-1885-5

Library of Congress Cataloging-in-Publication Data
Output measurements for medical ultrasound/Roy C. Preston, ed.
p. cm. Includes bibliographical references.
ISBN-13:978-1-4471-1885-5
1. Diagnosis, Ultrasonic—Equipment and supplies—Testing.
2. Ultrasonic waves—Measurement. I. Preston, Roy C., 1944–
RC78.7.U4095 1991
616.07'543—dc20 91-15413
CIP

Softcover reprint of the hardcover 1st edition 1991

Typeset by Photo·graphics, Honiton, Devon
69/3830-543210 Printed on acid-free paper

Preface

Medical applications of ultrasound range from therapy, foetal monitoring, diagnostic imaging and blood flow measurement, to surgery and lithotripsy. Ever since the earliest developments of medical ultrasound there has been an interest in the measurement and characterisation of the ultrasonic fields generated in these applications, either for assessment of performance of equipment or for safety. Until recently, procedures for acoustic output measurements were ill-defined, but standards are now being developed internationally by the International Electrotechnical Commission (IEC) and reliable commercial measuring equipment is becoming more widely available. The aim of this book is to introduce acoustic output measurement techniques and to provide an insight into the measurement methods.

The material presented is based on a course originally run during November 1989 at the National Physical Laboratory. In addition to the lectures, the course consisted of a series of practicals to allow participants to gain experience of making measurements.

Teddington, UK
February 1991

Roy C. Preston

Contents

7 Measurements on a Specific Acoustic Pulse Part 2: Measurements
D.R. Bacon

8 Real-Time Scanning Systems Part 1: Modes of Operation
A. Shaw

Contributors

D.R. Bacon, MA, PhD
Principal Scientific Officer

F. Davidson
Scientific Officer

R.C. Preston, BSc, PhD, C Phys, F Inst P
Senior Principal Scientific Officer

S.P. Robinson, BSc
Higher Scientific Officer

A. Shaw, MA
Higher Scientific Officer

B. Zeqiri, BSc, PhD
Senior Scientific Officer

All the above contributors are at the
National Physical Laboratory
Division of Radiation Science and Acoustics
Teddington
Middlesex TW11 OLW

R.C. Chivers, MA, PhD
Senior Lecturer
Physics Department
University of Surrey
Guildford
Surrey GU2 5XH

Symbols

a	Effective radius of a hydrophone active element
a_g	Geometrical radius of a hydrophone active element
a_1	Effective radius of an ultrasonic transducer
a_3, a_6	Effective radii of the active element of a hydrophone, determined from directional response measurements, at the -3 dB and -6 dB levels
a_{max}	Maximum effective radius for a specific hydrophone application
A_1	Effective area of an ultrasonic transducer
A_b	Beam-area
A_g	Geometrical area of an ultrasonic transducer
A_S	Scan-area
c	Speed of sound in a liquid (usually water)
C	End-of-cable capacitance of a hydrophone
C_{el}	Parallel input capacitance of an electrical load
f_{awf}	Acoustic working frequency
F	Radiation force on a target
F_g	$= 0.69A_g/(-6$ dB beam-area)
h	Target factor for radiation force balances
I	Instantaneous intensity
I_{bapa}	Beam-average pulse-average intensity
I_{bata}	Beam-average temporal-average intensity
I_m	Maximum intensity
I_{ob}	Output beam intensity
I_{pi}	Pulse-intensity integral
I_{sapa}	Spatial-average pulse-average intensity
I_{sata}	Spatial-average temporal-average intensity
I_{sppa}	Spatial-peak pulse-average intensity
I_{sppi}	Spatial-peak pulse-intensity integral
I_{spta}	Spatial-peak temporal-average intensity

I_{sptp}	Spatial-peak temporal-peak intensity
I_{ta}	Temporal-average intensity
k	$= 2\pi/\lambda$
l	Distance between hydrophone and transducer
l_1	Distance between an ultrasonic transducer face and the plane containing the spatial-peak temporal-peak acoustic pressure
L	Electrical loading correction for a hydrophone
M_c	End-of-cable open-circuit sensitivity of a hydrophone
M_l	End-of-cable loaded sensitivity of a hydrophone
p	Instantaneous acoustic pressure
p_{bar}	Beam-average rms acoustic pressure
p_{bap}	Beam-average pulse acoustic pressure
p_i	Pulse-pressure-squared integral
p_m	Mean-peak-cycle acoustic pressure
p_0	Temporal-peak acoustic pressure
p_p	Pulse acoustic pressure
prr	Pulse repetition rate
p_{sap}	Spatial-average pulse acoustic pressure
p_{sar}	Spatial-average rms acoustic pressure
p_{spp}	Spatial-peak pulse acoustic pressure
p_{spr}	Spatial-peak rms acoustic pressure
p_{sptp}	Spatial-peak temporal-peak acoustic pressure
p_{rms}	RMS acoustic pressure
p_+ (p_c)	Peak-positive acoustic pressure
p_- (p_r)	Peak-negative acoustic pressure
P or W	Temporal-average power
R	Radius of curvature of a focused transducer
R_{el}	Parallel input resistance of an electrical load
s_s	Separation between ultrasonic scan lines in the plane considered
S	Sensitivity factor for a radiation force balance
t_d	Pulse duration
U_L	End-of-cable voltage for a hydrophone
v	Instantaneous particle velocity
V_{rms}	RMS voltage
w_{pb6}	−6 dB pulse beam-width
w_{pb20}	−20 dB pulse beam-width
Z	Complex electrical impedance of a hydrophone
Z_{el}	Complex electrical input impedance of an electrical load
β	Nonlinearity parameter
θ_3, θ_6	Half angles for the directional response at −3 dB and −6 dB levels
λ	Acoustic wavelength in a liquid
ρ	Density of liquid (usually water)
σ_m	Nonlinear propagation parameter
ω	($2\pi f$) Angular frequency

Introduction

Roy C. Preston

Chapter 1 explains the importance of acoustic output measurements. Its format is somewhat different from the other chapters in that it sets out to answer questions which a clinician might have about acoustic output information. Although it is specifically aimed at the user of ultrasound during obstetric examination, the concepts are appropriate to other clinical applications of ultrasound. The chapter aims to answer a number of questions such as *'What is acoustic output information?'* and *'What type of acoustic output information is important?'* It also gives information about progress internationally on safety and standardisation and finally gives guidance on what the clinician should be doing to ensure continued safe use of equipment.

Chapter 2 starts by establishing the basic physics of ultrasonic wave propagation and then introduces the terminology used to describe propagation, such as particle velocity, acoustic pressure and intensity. Propagation through media is then considered by introducing the characteristic acoustic impedance, attenuation and absorption. This is followed by a section which deals with the generation of ultrasound using piezoelectric devices. Formulae are given to describe the acoustic fields generated by cylindrically symmetric piston sources. Finally, focused fields and nonlinear propagation are introduced as these are important features of real medical ultrasonic fields.

Chapter 3 considers the range of measurement methods which can be used to characterise medical ultrasonic fields. After introducing the key acoustical parameters, the next two sections deal with techniques for the determination of acoustic pressure and particle displacement respectively, both of which are dependent on the measurement position in the field. The fourth section deals with radiation force which can be considered either at a point in the field or integrated over the whole ultrasonic beam. The fifth section deals with temperature rise, which can also be considered locally or for the whole beam. The relative merits of the different techniques are then summarised and an attempt is made to compare the techniques on scientific, economic and ease-of-use terms. The basic conclusion is that the radiation force balance and the hydrophone have gained favour as the most suitable for the average laboratory or for industrial use. This chapter then introduces the concepts of traceability to national measurement standards and good measurement practice, a topic not dealt with very often in the field of ultrasonic measurements. Finally, a brief insight into qualitative assessment techniques is given.

The fourth chapter considers in detail the hydrophone as a measuring device. The different types of hydrophone in current usage are introduced with emphasis on the performance properties most relevant to their use for the characterisation of medical ultrasonic fields. The chapter goes on to discuss the problems associated with using hydrophones. Mounting the hydrophone, the measurement medium (water) and calibration are amongst the various topics dealt with. Finally, guidance on the choice of the most appropriate hydrophone is given.

Chapter 5 concentrates on the radiation force balance which is widely used as a means of measuring ultrasonic power. After introducing the basic measurement principle, the pros and cons of using an absorbing or reflecting target are discussed. This is followed by a general introduction to the various designs of radiation force balances and then three examples of radiation balances are described. To obtain the maximum benefit from using a radiation force balance, there are a number of precautions which need to be taken during measurements. These are dealt with in a systematic manner as are the treatment of a number of sources of systematic uncertainty in the measurement process.

Whilst four of the first five chapters deal with the measurement methods, the following chapters deal with the specification and measurement of ultrasonic fields using hydrophones. The first two deal with the measurement and characterisation of a single acoustic pulse and the following two deal with the same topics but for real-time scanners.

Chapter 6 deals with the characterisation and specification of ultrasonic fields, concentrating on the parameters which are currently referred to in a draft IEC and other standards. Since a hydrophone measures acoustic pressure, the primary parameters measured are pressure parameters and these are described in detail. It is usual to derive acoustic intensity parameters from the hydrophone measurements, and the range of parameters which can be derived in this way is given. Spatial-peak as well as spatial-average parameters are considered as is the derivation of total power by integrating the square of the hydrophone signal over the beam. In any real measurements, there is a need to locate maxima in fields and problems due to maxima of different acoustical parameters being located at different points are introduced. Finally, the concept of "*in-situ*" levels as required by the US Food and Drug Administration Reporting Requirements is considered together with its effect on the location of *in-situ* maxima.

Chapter 7 deals with the measurement process using a hydrophone to determine the various acoustical parameters for a specific acoustic pulse. Requirements for mounting the hydrophone and measuring the output voltage are discussed as are the bandwidth and spatial resolution requirements. Some simple rules are given which allow the most appropriate choice of hydrophone. Corrections to take account of electrical loading of the hydrophone output and the conversion of voltage to acoustic pressure and then intensity are considered. Continuous-wave fields are dealt with separately as these can pose different problems such as interfering reflections from tank walls or from hydrophones. Finally, an idea of the sources of measurement uncertainty and the overall accuracy of measurements is given for a whole range of acoustical parameters for measurements made using the NPL Beam Plotting Facility and the NPL Ultrasound Beam Calibrator.

The next two chapters extend the hydrophone measurement process to the characterisation of real-time scanning systems such as linear arrays, sector scanners etc. To do this it is necessary to understand the way a scanner works and this is dealt with in Chapter 8. The principles of operation of different types of real-time imaging system are considered in order to understand how the ultrasonic fields are likely to be distributed in space. The concept of overlapping lines is introduced as this is important for the reliable determination of the important parameter spatial-peak temporal-average intensity. The types of pulses generated by different modes of operation are illustrated and typical values for important acoustic output parameters given. In order to make output measurements on real-time scanners it is important to be able to recognise the pulse patterns observed using a hydrophone when a scanner is operating in different modes. To achieve this, typical pulse patterns for simple modes such as M-mode and B-mode are given, as are examples of more complex combined modes such as B+M. Multiple focal zones and repeated lines are also treated in the same way. Finally, the importance of understanding the front-panel settings of a scanner and the way they can affect acoustic output are considered with guidance being given on those settings which are most important.

Chapter 9 considers measurements for real-time scanners and the methods of coping with the complex fields. The first part deals with the measurement system requirements and concentrates on the need to obtain a reliable electrical trigger from the scanner. Various methods are presented together with techniques for using the trigger signals to synchronise the measurement system to the scanner pulsing sequence. The emphasis of the next section is on obtaining a repeating sequence so that one particular scan line can be selected. Different scan formats can introduce alignment problems and these are reviewed as is the need to consider overlapping scan lines when determining intensity. The effect of the overlap when dealing with multiple focal zone systems is illustrated as is the combination of multiple focal zones and *in-situ* levels. Finally, the determination of total ultrasonic power from measurements on single scan lines is discussed and measurement uncertainties are given at the end of the chapter.

Finally, a glossary is provided, giving an extensive list of definitions of parameters used in the field of acoustic output measurement. This list is rather extensive and is included as a reference for future use.

1

Safety of Diagnostic Ultrasonic Equipment: The Relevance of Acoustic Output Information to the Clinical User

R.C. Preston

1.1 Introduction

The purpose of this first chapter is to consider the relevance of acoustic output information to the clinical user. As obstetric examination is one of the most common uses of diagnostic ultrasound and one in which safety is of great importance, obstetric examination has been chosen as the main example to address the issue. In fact, many of the concepts presented here can be applied to other clinical applications of ultrasound such as physiotherapy and lithotripsy.

The safety of medical ultrasonic equipment used for obstetric examination is of concern to both patient and clinician. In assessing safety, an important consideration is the acoustic output of the equipment. This topic is highly complex for two main reasons. Firstly, the interaction of ultrasound with tissue is not fully understood, which means that it is not clear which of the many acoustic ways of describing the ultrasonic field generated by the equipment is most important. Secondly, the diagnostic equipment is complex, which means that it generates a variety of ultrasonic fields, thereby making the specification of the fields complicated.

To provide basic guidance on how to interpret acoustic output information, these notes are an attempt to provide simple answers to questions which the clinician might have on the safety of ultrasonic diagnostic equipment and to provide background information which could be of use in the future. The majority of the questions/answers apply equally to other applications of ultrasound in medicine such as physiotherapy and lithotripsy.

1.2 The Relevance of Acoustic Output Information

1.2.1 What Is Acoustic Output Information?

Acoustic output information is normally presented as a set of numeric values for specific parameters describing the ultrasonic field generated by the equipment, and these provide an indication of how much ultrasound is emitted by a scanner and transducer combination[1].

This information is normally given in terms of acoustic pressure, power, or acoustic intensity.

Acoustic pressure is a term which relates to the degree of compression and rarefaction experienced by a material when a sound wave travels through it. These compressions and rarefactions take place at the frequency of the sound, in our case at megahertz frequencies. The use of pressure is therefore a means of specifying the forces acting on the material through which the ultrasound passes.

An ultrasonic wave has the capability of transferring energy as it propagates. This is termed the power of the beam, and is a function of the total beam.

Beam intensity is normally specified at a particular point in the field and is given in terms of power passing through a unit area in a particular direction.

1.2.2 What Is Acoustic Output Information Used For?

Acoustic output information gives the technical expert information about the ultrasonic field which can then be used for assessment of performance. It is therefore relevant to the specification of equipment performance. It also gives guidance on safety; the higher the output the greater the theoretical risk.

1.2.3 Why Should You Know the Acoustic Output Levels of Your Equipment?

A knowledge of output level allows you to make a relative risk assessment of different modes of operation or of different pieces of equipment. It therefore allows you to satisfy yourself and your patient that the use of the equipment is "safe".

1.2.4 Are Acoustic Output Levels Really that Important?

New knowledge about the interaction of ultrasound with tissue will in future lead to better understanding of safety issues and to changes in diagnostic techniques and good clinical practice.

The situation is always changing. For instance, new equipment developments can lead to higher (sometimes lower) acoustic output levels as can upgrades and new transducers. New modes of operation can affect the output levels (e.g. colour Doppler).

1.3 Acoustic Output Information

1.3.1 What Type of Acoustic Output Information Is Important?

A very large number of parameters has been defined to describe the ultrasonic field but the following group is believed to represent those of greatest importance:

p_+ and p_-: Peak pressures
The largest positive and negative acoustic pressures in the field. These usually occur at or near the focal point i.e. the point in the field at which you expect the best image to be obtained.

I_{spta}: Spatial-peak temporal-average intensity
The highest intensity in the ultrasonic field, again expected at or near the focal point.

W: Power
The total acoustic power in the beam. Total power does not depend on position in the field – assuming all the beam is included in the determination.

I_{ob}: Output beam intensity
The power divided by the radiating area of the transducer.

w: Beam width
The width of the ultrasonic beam at the focus.

In practice, it is normally necessary to give other information in order to relate the acoustic output levels to equipment operating conditions. For instance, the mode of operation (B, M or D) is crucial and so are the front panel settings. A well-documented specification of acoustic output will give sufficient information to enable the scanner to be set up to reproduce the conditions which correspond to the specified output levels. Unfortunately, this is not always the case – manufacturers provide some information but not in a common format.

Although measurements are in practice made with the ultrasound propagating in water, the specification of acoustic output by manufacturers can sometimes be in terms of *in-situ* values (see 6.12). This approach stems from the attempt by the United States Food and Drug Administration (FDA) [2] to estimate the acoustic levels in tissue. These *in-situ* levels are lower than the water values because of the higher attenuation in tissue. Consequently, care must be exercised when comparing numbers from different sources as they may represent different types of data. Unfortunately, FDA regulates acoustic output on the basis of *in-situ* values and hence they are often quoted.

1.3.2 What Type of Information Is Given?

It is normal to specify the maximum acoustic output levels. Although these may be the levels used in many clinical examinations, front-panel settings will often be used which yield output levels below the maximum. If the system can produce high levels (say I_{spta} greater than 100 mW/cm^2), it is important to know how these relate to the output levels used in typical clinical examinations. This information should be available from the manufacturer.

Some equipment provides on-screen indication of intensities. As the output levels are changed during clinical examination, so the indications of intensities also change giving the operator direct feedback of exposure information.

1.3.3 Why Are Different Parameters Important?

Acoustic pressure is related to the forces that act on molecules as the acoustical vibrations pass through the medium. Hence, as the vibrations represent compressive and rarefactive forces, the larger the amplitude of vibration the larger are the forces trying to pull the molecules apart. In certain situations of excessive force, cavitation and other non-thermal effects can occur.

Power and intensity parameters relate to how much heat could be produced when the ultrasound passes through an absorbing medium. Water and fluids are low attenuators and would not be significantly heated as the ultrasound passes through them. Soft tissue absorbs about 10% to 20% of the power per centimetre path and bone (adult) absorbs it all in a very short distance. Foetal bone is known to have an attenuation which increases with foetal age. Consequently, heating would be minimal in fluids and maximal in adult bone. (Absorption also affects our ability to image through tissue although scattering is also of major importance in this case.)

The ultrasonic beam width is important because it gives information about the area over which interaction of the beam with tissue takes place.

1.3.4 How Can You Obtain Acoustic Output Information?

There are two main sources of acoustic output information:

1. *From manufacturers*. Manufacturers should be able to provide information and this is the preferred route. Information should be provided in the equipment manuals. As already mentioned, on-screen indication of intensities is provided on some products.

2. *Have measurements made*. A number of laboratories and test houses can undertake measurements and international standards are now being developed to form the basis of such measurements but there is, as yet, no formal accreditation of such laboratories. In due course, there may be a structure of secondary standardising laboratories. In the UK, such laboratories would be capable of undertaking measurements traceable to NPL, the UK's national measurement standards laboratory. In the UK, NPL provides a measurement service for medical ultrasonic equipment.

1.3.5 Are There Sources of Published Data on Acoustic Output?

Two types of acoustic output information are published. The first is statistical information giving the ranges of output levels for different equipment but this type of publication does not give the product names. More useful recent sources of information relating to named products are:

1. AIUM Publication. Acoustical data for diagnostic ultrasound equipment 1987. American Institute of Ultrasound in Medicine, Rockville, Maryland 20852-3139, USA.
2. Series of reports produced by the Joint UK Health Departments, Ultrasound Equipment Evaluation Project, available from the Department of Health, Russell Square, London, UK.

1.4 Acoustic Fields and Output of Medical Ultrasonic Equipment

1.4.1 What Types of Ultrasonic Fields Are Generated?

The following sections give an idea of the clinical applications of the different types of medical ultrasonic equipment together with a description of the various types of fields they generate. Finally, the range of acoustic output levels are given for the different types of equipment.

1.4.2 What Type of Field Is Generated by Diagnostic Imaging Equipment?

Diagnostic imaging using ultrasound finds applications in all soft tissues providing gas and bone are avoided, one of the main applications being the abdomen (obstetrics). A wide range of different probes operating at different frequencies is available, for penetration to different depths and for imaging different tissue slices. Information is gathered in B- or M-mode (also combined). B-mode is a two-dimensional cross-section (as seen on the viewing screen) which means that the acoustic energy is distributed in a plane as the beams are fired sequentially in a fan (sector scanner), rectilinear (linear array) or any other format. M-mode is a display of the signals received from one scan line as a function of time (depth versus time display on the viewing screen). Pulsed excitation of the transducer is used for both B- and M-modes and various methods are employed to steer the beam such as:

Mechanically scanned single-element

Mechanically scanned annular array

Phased array

Because of the need to obtain as much fine detail as possible during the imaging process, the acoustic beams are focused which means that they tend to give rise to high acoustic pressures.

1.4.3 What Type of Field Is Generated by Doppler Equipment?

Pulsed Doppler.
Clinical applications of pulsed Doppler include cardiovascular and peripheral vascular. The principle of operation is that a frequency shift of the scattered or reflected ultrasound occurs if it is incident on a moving target. In general, longer pulse lengths are needed than in pulse-echo in order to define the frequency. Various geometries are used to steer and focus the beam (similar to the diagnostic imaging case), but pulsed Doppler is often used with the beam stationary (single orientation), which increases the local exposure levels. Colour-flow mapping systems often use the Doppler type of acoustic pulse but scanned in a similar way to imaging techniques. Almost every manufacturer uses different methods; some combine imaging and Doppler pulses, others use the same type of pulse for both purposes. However, despite the large variation in systems, they all generate sequences of acoustic pulses. Acoustic pressure levels may be comparable with those used in diagnostic imaging but the time-average intensities and powers can be significantly greater because the pulse lengths are longer and pulse repetition rates often higher.

Continuous Wave.
Clinical applications of continuous wave Doppler are similar to those of pulsed Doppler but no depth information is available. A double transducer geometry is normally used, mainly with D-shaped elements (transmit and receive). The focusing is usually achieved using a lens. The acoustic pressures are lower than those for pulsed Doppler but the time-average intensities may still be high. Doppler systems used for foetal monitoring often employ large transducers generating broad and sometimes diverging beams. The acoustic pressures are usually relatively low with the maximum values near the face of the transducer. Total powers, however, may not be that low.

1.4.4 What Type of Field Is Generated by Physiotherapy Equipment?

Clinical applications of ultrasound physiotherapy are widespread. The therapeutic effect is not well understood but is based in part on tissue heating. While large transducers are frequently used, small applicators are popular for special tasks. The transducers are mainly plane piston sources – air-backed and hence highly resonant and efficient to enable high powers to be generated. They are usually operated in continuous wave (cw) or long tone-burst mode. Acoustic pressures are typically 0.5 MPa and intensities can be up to a few watts per square centimetre although most applications use less than 0.5 W/cm^2.

Transducers are usually non-focused but can have a horn or a lens attached to the front surface.

1.4.5 What Type of Field Is Generated by a Lithotripter?

Lithotripters are not normally considered as ultrasonic devices but in practice the acoustic fields they generate are not too dissimilar from those of diagnostic systems. In general, the acoustic pulse duration is longer and the acoustic pressure is larger. Different methods of generating and focusing the fields are employed. Some use a focal bowl covered with a large number of piezoelectric elements, while others use an elliptical bowl with a spark source at one focus. The destructive action is essentially a mechanical one and is the subject of current research; cavitation plays an important role.

1.4.6 What Are the Acoustic Output Levels of Different Equipment?

The ranges of values for some of the important parameters and characteristics of the different medical ultrasonic fields are given in Table 1.1.

Table 1.1. Characteristics of medical ultrasonic fields

	Diagnostic	Pulsed Doppler	Therapy	Lithotripter
Pressure (MPa)	1–8	1–5	0.1–0.7	10–100
I_{spta} (mW/cm^2)	1–30	300–5000	20–10 000	10–100
Total power (mW)	1–100	20–250	100–15 000	1–50
Frequency (MHz)	2–10	2–10	0.75–3.0	0.2–0.9
Pulse duration (μs)	0.2–1.0	0.3–10	cw or ms pulse	0.2–10

1.5 Standards and Regulations

1.5.1 What Is Being Done Internationally About Acoustic Output and Safety?

The International Electrotechnical Commission (IEC) has two committees active in this field. Standards produced by the IEC invariably become European Standards through Comité Européen de Normalisation Electrotechnique (CENELEC) and British Standards (BS).

IEC Sub-committee SC62D: Safety of Electromedical Equipment, is producing a standard dealing with requirements for safety of ultrasonic diagnostic equipment, both electrical and acoustical. The acoustic output requirements will be based on work of IEC Technical Committee 87: Ultrasonics.

IEC Technical Committee 87 has produced [1] a draft International Standard entitled "Requirements for the declaration of the acoustic output of medical diagnostic ultrasonic equipment" which will require manufacturers to give information on acoustic output in a well-defined manner. It also specifies a set of exemption levels, defined as the acoustic output levels below which a manufacturer need not give detailed acoustic output information. By implication equipment generating levels below these specified levels is considered unquestionably safe. The exemption levels proposed are:

Peak-negative acoustic pressure, p_-	1 MPa
Spatial-peak temporal-average intensity, I_{spta}	100 mW/cm^2
Output beam intensity, I_{ob}	20 mW/cm^2

This IEC committee is also mindful of the need for recommendations on maximum acoustic levels used in different clinical applications and is currently exploring the route of producing a classification standard. Such a standard would provide the clinical user with a simple assessment of the risk of using different transducers and fields.

The World Federation for Ultrasound in Medicine and Biology (WFUMB) has a Safety and Standardisation committee and is trying to generate a consensus of the various national statements and documents. WFUMB is currently producing a report on thermal effects of ultrasound.

1.5.2 What Is Being Done Nationally About Acoustic Output and Safety?

Considerable effort has been expended in the USA, both at a Federal level and at a professional and advisory level. The US FDA has placed reporting requirements [2] on manufacturers which ensure that equipment is not used for a particular clinical application if it generates higher acoustic output levels than similar equipment produced for the same application prior to the particular Federal Law Enactment in 1976. More recently, FDA has encouraged the display of output information on the viewing screen and systems are now on the market with this feature. Discussions are currently taking place on alternative methods of providing acoustic output information to the clinician and on-screen labelling of information more directly relevant to safety such as thermal and cavitation (or mechanical) indices are being considered.

The AIUM in conjunction with the National Electrical Manufacturers Association (NEMA) produced a safety standard in 1981 [3] which dealt with the specification of acoustic output and made a number of safety statements. One of the most important of these was a re-confirmation of an earlier statement, namely:

> In the low megahertz frequency range there have been, as of October 1978, no confirmed significant biological effects in mammalian tissues exposed to intensities[a] below 100 mW/cm^2. Furthermore, for ultrasound exposure times[b] less than 500 seconds and greater than one second, such effects have not been demonstrated even at higher intensities, when the product of intensity and exposure time is less than 50 joules/cm^2.
>
> [a]Spatial-peak temporal-average intensity.
> [b]Exposure time includes off-time as well as on-time for repeated-pulse regimes.

More recently [4], a new safety statement has been produced by AIUM:

> In the low megahertz frequency range there have been (as of this date) no independently confirmed significant biological effects in mammalian tissues exposed in-vivo to unfocused ultrasound with intensities[a] below 100 mW/cm^2, or to focused ultrasound[b] with intensities below 1 W/cm^2. Furthermore, for exposure times[c] greater than 1 second and less than 500 seconds (for unfocused ultrasound) or 50 seconds (for focused ultrasound) such effects have not been demonstrated even at higher intensities, when the product of intensity and exposure time is less than 50 joules/cm^2.
>
> [a]Spatial-peak temporal-average intensity.
> [b]Quarter power beam width smaller than four wavelengths or 4 mm, whichever is less at the exposure frequency.
> [c]Exposure time includes off-time as well as on-time for repeated pulse regimes.

This more recent statement accepts that focused fields may create higher intensities. The reason for the increase in the intensity from 0.1 to 1 W/cm^2 for focused fields is related to the belief that, when considering thermal effects of ultrasound, heating will be less severe in focused fields. The concept of predicting heating is being taken further by a number of national bodies. One of the most important is the National Council for Radiation Protection and Measurements [5,6] in the USA which is producing an advisory document on thermal effects of ultrasound. This will give information to enable the clinician to estimate the temperature elevation during ultrasound examination. The concept of tissue models is being introduced, accepting that different clinical applications of ultrasound could lead to different heating because of the different tissue attenuations. The three trimesters in obstetrics are considered and different attenuations are proposed for each.

In Europe there have been a number of special meetings of experts and in some cases committees have produced reports. For instance, the Royal College of Obstetricians and Gynaecologists [7] produced a report on Routine Ultrasound Examination in Pregnancy with a recommendation that manufacturers should be required to publish acoustic output information. Such information would allow clinicians to implement a recommendation in a report entitled "The safety of diagnostic ultrasound" [8] produced by the British Institute of Radiology which recommended that "exposure conditions are adequately documented". The European Federation of Societies for Ultrasound in Medicine and Biology has for some years supported a "watchdog" group which reviews published scientific papers on biological effects of ultrasound.

Despite a large amount of effort being expended on the topic, there are as yet no requirements in Europe for either the declaration of acoustic output information or setting maximum levels. This will however change in the future with the publication of IEC standards.

1.6 The Impact on the Clinician

1.6.1 What Should You Be Doing?

There are three sensible precautions you can take:

1. Ensure that you know the acoustic output levels of your equipment, or at least that you have ready access to the information.

2. Know which modes give the highest output levels; highest pressures, intensities and powers.
3. When guidance on prediction of worst-case temperature elevations is published, make sure your equipment cannot produce temperature increases of more than 1°C. If it can, make sure you know which mode/modes exceed 1°C so that you can take greater care when using them clinically.

1.6.2 What Are the Consequences of High Power/Intensities?

As a consequence of high intensities, the probe surface may get hot (self-heating), or heating of attenuating tissue could take place.

1.6.3 What Modes Produce the Highest Power/Intensities?

Static Doppler beams usually produce the highest intensities and powers. M-mode can also be high as can some types of colour-flow mapping systems. The lowest I_{spta} levels are usually obtained with imaging systems during simple B-mode imaging. However, even in B-mode, the total power from the probe may be high (distributed over the sector or rectilinear beam) and this can lead to temperature increases near or at the probe surface from self-heating of the probe.

1.6.4 What Intensity Levels Are Acceptable?

There is no simple answer to this question. The objective should be to ensure that the intensities and powers are such that there is no risk of increasing the temperature of any part of the patient during ultrasound examination by more than 1°C. This temperature increase is chosen for a number of reasons. First, studies have shown that 1°C increase should be acceptable even for the most sensitive organs in the body and for tissues at their most sensitive stages of development such as during early pregnancy. Second, it is generally accepted that any unintentional increase in temperature resulting from exposure to other non-ionising radiation fields should not exceed 1°C. Eventually, there will be reliable methods of predicting the temperature increase, promulgated by bodies such as NCRP and IEC as noted earlier. In the meantime, guidance can be obtained from the safety statements given earlier and from existing regulatory practice in the USA and Japan (not in Europe). For obstetric examination, the US FDA [2] sets the following upper levels:

	In water	*In-situ*
I_{spta} (mW/cm^2)	180	94
I_{sppa} (W/cm^2)	350	190
I_m (W/cm^2)	550	310

Note: Values for spatial-peak pulse-average intensity, I_{sppa}, and maximum intensity, I_m, are given above for completeness but will not be discussed here.

The limit of 180 mW/cm^2 for I_{spta} is clearly similar to the 100 mW/cm^2 statement by AIUM [3] and is close to the figure of 240 mW/cm^2 given by Japanese Industrial Standards [9]. These are taken to be the levels below which there are no proven adverse biological effects. As mentioned earlier, the IEC has a draft International Standard which sets levels below which the equipment can be considered safe. The levels of 100 mW/cm^2 for I_{spta} and 20 mW/cm^2 for the output beam intensity, I_{ob}, are specified.

Consequently, to be as safe as possible, the maximum level of 100 mW/cm^2 for I_{spta} should be adopted, and in a recent article in the Bulletin of the British Medical Ultrasound Society [10], the 100 mW/cm^2 limit was recommended.

1.6.5 What Can You Do to Minimise Exposure of Patients to High Power/Intensities?

Four precautions should be taken to minimise patient exposure:

1. Ensure the equipment produces no acoustic output when in frame freeze.
2. Never keep the probe stationary on the patient for longer than is necessary to obtain adequate clinical information.
3. If you change front-panel settings such as pulse repetition rate, output level, gate length and range-gated depth, use those settings which are known to produce lower outputs and which at the same time maintain an adequate level of diagnostic information. This means reducing the pulse repetition rates and output levels and using the shortest gate lengths. It is essential to build up knowledge about the best front-panel settings of the equipment for different clinical applications. In conclusion, "get to know your equipment".
   Unfortunately, many Doppler systems automatically alter the system settings to maintain constant I_{spta}. You should insist that manufacturers give you information on how the acoustic output depends on key front-panel settings.
4. Ensure that the probe surface temperature does not increase significantly, especially for intra-cavity probes such as those used in obstetrics and gynaecology.

1.6.6 What Are the Consequences of High Acoustic Pressures?

High acoustic pressures may produce cavitation which can be harmful. Cavitation is a complex subject but it is believed that harmful effects can occur if bubble activity leads to collapsed transient cavitation events. Such events can produce local high temperatures and pressures and be the source of chemical free-radicals. Peak-negative acoustic pressure (or the amplitude at the fundamental frequency of the pulse) is believed to be the most relevant parameter when considering the likelihood of cavitation taking place.

1.6.7 What Modes Produce the Highest Pressures?

Diagnostic imaging (B-mode) usually gives the highest pressures. However, M-mode often uses the same acoustic pulse as do many of the combined modes, such as those with Doppler. In general, cavitation thresholds are lower for lower acoustic frequencies. This means that cavitation is less likely to occur when using a 7.5 MHz probe than when using a 2 MHz probe.

1.6.8 What Pressure Levels Are Acceptable?

This is an even more difficult question to answer than the equivalent one on intensities. Current studies indicate that there are unlikely to be any adverse effects from cavitation at the levels used in medical diagnosis. Nevertheless, peak pressures from diagnostic equipment are very high. Peak-negative pressures can be over 4 MPa and cavitation thresholds in water are less than 1 MPa. However, tissue attenuation and the greater difficulty of sustaining cavitation in tissue are two factors which mean that cavitation may not be taking place in tissue during diagnostic examinations.

No regulatory requirements currently exist which explicitly limit the maximum acoustic pressures from diagnostic equipment. The nearest to a limit is that given by the FDA [2] for the parameter I_m, the maximum intensity during the largest half cycle. By making assumptions about the ratio of the positive and negative pressures, the FDA regulation can be shown to be equivalent to a peak-negative acoustic pressure of about 2 MPa. Many diagnostic systems generate acoustic pressures similar to or above this level. The article in the BMUS Bulletin [10] suggested a maximum level of 3.8 MPa (positive and negative). Whilst this represents the upper limit of negative acoustic pressures generated by modern diagnostic equipment, positive acoustic pressures are often greater.

The IEC draft International Standard [1] puts a level of 1 MPa on the peak-negative acoustic pressure below which manufacturers need not provide detailed acoustic output information. By implication this sets a level below which the equipment is considered completely safe.

1.6.9 What Can You Do to Minimise Exposure to High Acoustic Pressures?

Use the lowest output settings capable of producing adequate clinical information.

1.6.10 What Should You Do to Keep Up to Date?

Performance and safety standards of equipment are subject to periodic update.

1. Make sure you know the acoustic output levels of your equipment
2. Ensure you know how any new transducers or upgrades have affected the acoustic output

3. Take note of guidelines on safety produced by bodies such as IEC, WFUMB, World Health Organization (WHO), European Federation of Societies for Ultrasound in Medicine and Biology (EFUMB) (Watchdog Group), BMUS, NCRP and AIUM

1.6.11 What Should You Do if You Have the Opportunity of Buying New Equipment?

In evaluating new equipment, consider the following:

1. Ensure manufacturers provide reliable acoustic output information
2. Ask manufacturers for details of how the information they provide is obtained
3. Choose equipment which, all other things being equal, has the lowest output levels
4. Pay particular attention to the output of Doppler modes

1.7 Conclusions

In summary, these points are important:

- Know the acoustic output levels of your scanner
- Know how the levels compare with the IEC exemption levels, the AIUM safety statements and the FDA regulatory levels
- Know how the acoustic output levels of your scanner vary with system operating modes and front-panel settings
- Insist on manufacturers providing acoustic output information
- Use minimum exposure levels consistent with obtaining adequate clinical information
- Take all opportunities to reduce exposure if the IEC, AIUM or FDA levels are exceeded

1.8 Recommended Additional Reading Material

Kremkau FW. Diagnostic ultrasound: Principles, instrumentation and exercises. Grune and Stratton Inc., Orlando, Florida, USA, 1984

Hill CR (ed). Physical principles of medical ultrasound. Ellis Horwood series in applied physics. Ellis Horwood Ltd, Chichester, UK, 1986

Nyborg WL and Ziskin MC. Biological effects of ultrasound. Churchill Livingstone, New York, 1985

Wells PNT. Biomedical ultrasound. Academic Press, London, 1977

WHO. Ultrasound. Environmental Health Criteria 22. World Health Organization, Geneva, Switzerland, 1982

Williams AR. Ultrasound: biological effects and potential hazards. Academic Press, London, 1983

1.9 References

1. IEC. Draft International Standard: Requirements for the declaration of the acoustic output of medical diagnostic ultrasonic equipment. Document 87(CO)11, International Electrotechnical Commission, Geneva, Switzerland, 1990
2. FDA. Guide for measuring and reporting acoustic output of diagnostic ultrasound medical devices. Document 510(k), US Department of Health and Human Services, Food and Drug Administration, HFZ-132, Rockville, MD 20857, USA, 1985
3. AIUM/NEMA. Safety standard for diagnostic ultrasound equipment. Publication UL 1-1981. National Electrical Manufacturers Association, 2101 L Street NW, Washington DC 20037, USA, 1981
4. AIUM. Bioeffects considerations for the safety of diagnostic ultrasound. J Ultrasound Med 1988; 7 No. 9 (Supplement)
5. NCRP. Biological effects of ultrasound: mechanisms and clinical implications. Report No. 74, National Council on Radiation Protection and Measurements, Bethesda, MD 20814, USA, 1983
6. NCRP. Exposure criteria for medical diagnostic ultrasound. Part 1. Criteria based on thermal mechanisms. Draft Report of Scientific Committee No. 66, National Council on Radiation Protection and Measurements, Bethesda, MD 20814, USA, 1990
7. RCOG. Report of the RCOG working party on routine ultrasound examination in pregnancy. Royal College of Obstetricians and Gynaecologists, 27 Sussex Place, Regents Park, London NW1 4RG, 1984
8. BIR. The safety of diagnostic ultrasound. Wells PNT (ed), British Journal of Radiology Supplement No. 20, The British Institution of Radiology, London, 1987
9. Maeda K, Ide M. The limitation of the ultrasound intensity of diagnostic devices in the Japanese industrial standards. IEEE Trans Ultrason Ferroelec Freq Contr 1986; UFFC-33:241–243
10. BMUS. Prudent use of diagnostic ultrasound. British Medical Ultrasound Society Bulletin No. 50, British Medical Ultrasound Society, London, 1988

2

Fundamentals of Ultrasonic Propagation

R.C. CHIVERS

2.1 Introduction

The purpose of this chapter is primarily to provide an insight into the fundamentals of ultrasonic wave propagation and, thereby, give an introduction to some of the physics which underpins the basic measurement methods employed to determine the acoustic output of medical ultrasonic equipment. Many of the characteristic parameters and much of the basic terminology will be introduced here and taken further during later chapters. However, these later chapters will deal with the practical aspects of measurement and characterisation, often assuming that the basic physical principles of measurement have been established.

2.2 Ultrasonic Wave Propagation

2.2.1 Types of Wave

Ultrasonic waves are mechanical waves with a frequency above the limit of human hearing, typically 20 kHz. There is a fundamental higher limit of frequency based on the closest atomic spacing in solids and the Nyquist sampling theorem of about 10^{13} Hz.

The fundamental parameter is the displacement of a small part of the material. It is tempting to identify this with an atom or molecule but the concepts of wave propagation assume a continuum so that wave descriptions are concerned with a distance scale large compared with many atoms or molecules. Thus formally we refer to "the particle". Hence, the particle displacement ξ which represents the instantaneous displacement of the particle from its equilibrium position, is the quantity of interest. There are two main classes of waves – longitudinal and transverse in which the particle displacement is respectively parallel to and perpendicular to the direction in which the wave propagates (Fig. 2.1). These waves are often called compressional and shear waves respectively. The former may propagate in any medium, the latter only propagate in solids. This is because the weaker bonds between atoms and molecules in liquids and soft tissues do not readily allow transverse displacements to be transferred to an adjacent point in the direction of propagation. In addition to these two types of wave, surface (or Rayleigh) waves may propagate at the boundary of a liquid with a solid; and plate (or Lamb) waves if the dimensions of the medium are comparable with the wavelength.

2.2.2 Wave Parameters

The particle displacement, ξ, can – in the absence of losses – be shown to satisfy the linear wave equation for small amplitude waves, whether the waves are longitudinal or transverse:

$$\nabla^2 \xi = \frac{1}{c^2} \frac{\partial^2 \xi}{\partial^2 t}$$

where c is the speed at which the waves travel.

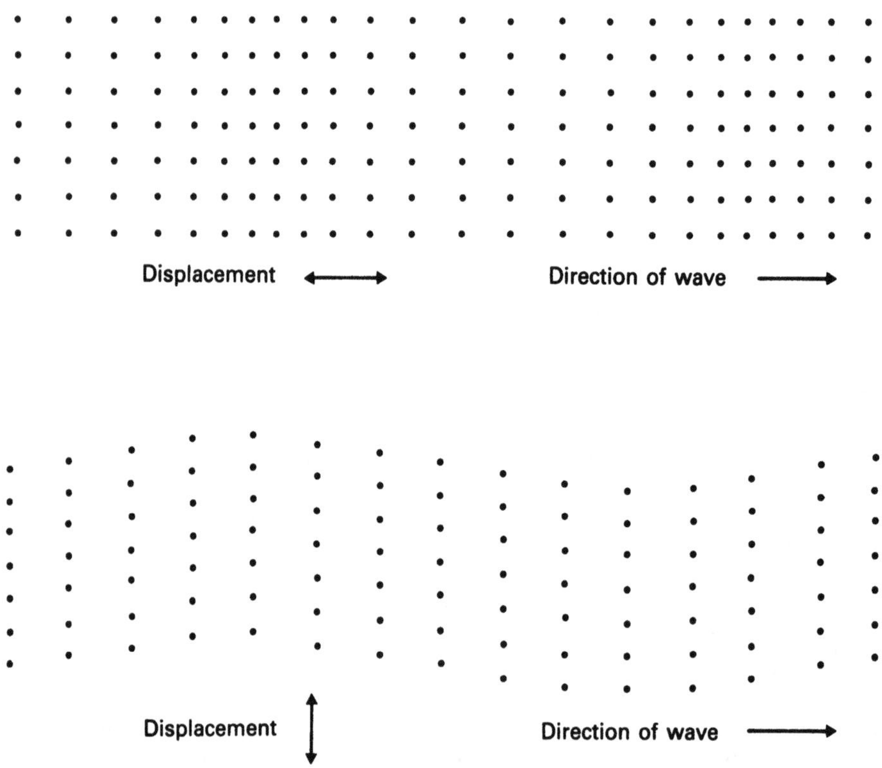

Figure 2.1. Particle displacements for longitudinal (compressional) and transverse (shear) waves.

If the particle displacement satisfies the wave equation, so also will the particle velocity $\dot{\xi}$ and the particle acceleration $\ddot{\xi}$. As a compressional wave propagates, the relative displacement of different parts of the medium leads to localised changes of density. These are described by the condensation, s, which also satisfies the wave equation and is defined by:

$$s = \frac{\rho - \rho_0}{\rho_0}$$

where $\rho = \rho(x,t)$ is the local density and ρ_0 the equilibrium density.

Localised changes of density cause localised changes of pressure, p, (referred to as acoustic pressure) through the equation of state for the medium:

$$p = P - P_0 = As + \frac{Bs^2}{2} + \ldots \qquad (2.1)$$

where P is the absolute pressure, P_0 is the ambient pressure and A, B are constants. If only the first term is taken we obtain the linear wave equation which the acoustic pressure, p, also satisfies. In some circumstances (including diagnostic devices) it is necessary to consider nonlinear waves (see 2.5). It is interesting to note that while $p = P - P_0$ cannot be negative in a gas, it can be in liquids or solids. This simply describes a situation in which the stress over any plane in the material is tensile (tending to pull it apart) rather than compressive. Both liquids and solids can support large tensile stresses before they rupture. Thus rupturing, in liquids, is associated with the term cavitation.

In addition, there is another parameter which satisfies the linear wave equation, the velocity potential, ϕ. While it has no physical identity *per se*, it provides measurable physical quantities, through derivatives:

$$\underline{\dot{\xi}} = -\underline{\nabla}\phi$$

and

$$p = \rho_0 \frac{\partial \phi}{\partial t}$$

Note: The bar denotes a vector quantity.

In many respects particle velocity and acoustic pressure play analogous roles in acoustics to those played respectively by current and voltage in AC circuit theory.

2.2.3 Linear Wave Propagation

The one-dimensional solution to the linear wave equation is of the form (for sinusoidal pressure waves):

$$p = p_1 \sin(\omega t - kx) + p_2 \sin(\omega t + kx)$$

which gives a sinusoidal variation with x at fixed t and with t at fixed x. The first term represents a wave travelling to the right, the second one a wave travelling to the left. It may be written in exponential form:

$$p = p_1' \exp i(\omega t - kx) + p_2' \exp i(\omega t + kx)$$

Here $\omega = 2\pi f$ is the angular frequency, f is the frequency, and $k = 2\pi/\lambda$ is the wave number, λ is the wavelength. Wavelength and frequency are related by:

$$c = f\lambda = \frac{\omega}{k}$$

The speed at which the waves travel can be shown to be determined by the choice of medium. For compressional waves:

$$c^2 = (\rho_0 m_0)^{-1}$$

when m_0 is the compressibility of the material. Once the material is chosen, c is fixed. We may then vary f to obtain a suitable value for λ for the application in mind. In medical ultrasound, the need to image structures with dimensions of the order of millimetres requires the use of wavelengths of similar magnitude. As the speed of sound in soft tissues is typically 1.5 mm/μs, this requires the use of megahertz frequencies. In order to image smaller features, higher frequencies are used, for instance skin scanners use frequencies above 20 MHz.

Care must be taken to note that the phenomenon of wave propagation involves three dimensions in general. Whereas many textbooks use a one-dimensional approach for simplicity (as indeed we shall) it can lead to omission of some important physical features. Most important of these is that some of the quantities we have discussed in 2.2.2 are actually vectors, specifically ξ and its derivatives. For a one-dimensional longitudinal wave their direction is defined as being in the direction of wave propagation and therefore the explicit vector notation has been dropped from now on.

The acoustic pressure, p, and the particle velocity, $\dot{\xi}$, can be combined to yield two other quantities of importance. Their ratio $p/\dot{\xi} = Z$ gives the acoustic impedance. (The AC circuit theory analogue is electrical impedance.) In general, this is complex and depends on the shape of the wave. Thus, for spherical waves it is of the form:

$$Z = \rho_0 c \left(1 + \frac{iA}{r}\right)$$

where A is a constant and r is the radial coordinate. For very large r the last term vanishes, Z is real and is called (for plane waves) the characteristic acoustic impedance. This is a property of the medium only. Typical values for selected materials are given in Table 2.1.

The product of p and $\underline{\dot{\xi}}$ (time averaged) is the intensity, I, of the wave:

$$\underline{I} = \langle p \underline{\dot{\xi}} \rangle$$

Note that it is a vector quantity and represents the energy flowing per unit time through unit (vector) area. As discussed in Chapter 1, forms of intensity are usually used as the basis for discussions of exposure.

If we have plane waves:

$$I = \frac{1}{2} Z \dot{\xi}_0^2 = \frac{1}{2} \frac{p_0^2}{Z}$$

Table 2.1. Values of speed of sound and characteristic impedance of different materials

Material	Speed of sound (m/s) c	Characteristic acoustic impedance ($kg/m^2/s \times 10^6$) Z
Air (STP)	331	4×10^{-4}
Water	1480	1.48
Castor oil	1500	1.43
Perspex	2680	3.20
Aluminium	6400	18.0
Fat	1450	1.38
Brain	1541	1.58
Liver	1549	1.65
Blood	1570	1.61
Skull bone	4080	7.80

where $\dot{\xi}_0$ and p_0 are the amplitudes of the sinusoidal waves. (Note the parallel with AC circuit theory.) Table 2.2 gives values for different acoustical parameters at different intensities, both in water and in air. It is noticeable that the condensation, s, for air is more than two orders of magnitude greater than that for water, reflecting the greater compressibility of gases compared to liquids.

If we have spherical waves the intensity decreases in proportion to the square of the distance from the origin of the spherical waves. This is relevant for the finite transducers we use in instrumentation (see 2.3).

Provided the waves are of small enough amplitude to obey the linear wave equation, there is no loss of generality by using sinusoidal dependencies for the waves since any arbitrary shape of pulse in the time domain can be constructed by a linear superposition of sinusoids using a Fourier transform.

The energy density, ϵ (the energy per unit volume) is related to the intensity by:

$$I = \epsilon c$$

The integration of I over an area such as that of a transducer gives the total power passing through that area:

$$W = \int I \, \mathrm{d}S$$

Table 2.2. Values of different acoustical parameter amplitudes at 1 MHz

Intensity (W/cm^2)	ξ (cm)	$\dot{\xi}$ (cm/s)	$\ddot{\xi}$ (cm/s^2)	s	p (atm)
Water					
0.01	0.183×10^{-6}	1.15	7.22×10^6	0.762×10^{-5}	0.171
1	1.83×10^{-6}	11.5	72.2×10^6	7.62×10^{-5}	1.71
100	18.3×10^{-6}	115.0	722×10^6	76.2×10^{-5}	17.1
Air					
1.0	11.2×10^{-5}	700	44×10^8	20.1×10^{-3}	28.1×10^{-3}

If a target intercepts an ultrasonic beam, the target experiences a force known as the "radiation force". For a plane wave incident on a perfectly absorbing target, the radiation force is given by:

$$F = \frac{W}{c} = \frac{IA}{c} = \epsilon A$$

where A is the area of the beam intercepted.

If the target is planar and with its normal making an angle α to the direction of the ultrasonic beam, the force in the direction of the incident beam is given by:

$$F = \frac{W}{c}(1 + R\cos 2\alpha)$$

where R is the intensity reflection coefficient (see 2.2.4) of the surface and it is assumed that all radiation not reflected is absorbed by the target. The measurement of radiation force is one of the important methods of ultrasonic output assessment (see Chapters 3 and 5).

2.2.4 Mechanics of Echo Production

For linear waves, the principle of superposition applies (i.e. the result of two or more overlapping waves is simply the sum of the waves). In this case, any shape of wavefront can be obtained from a combination of plane waves travelling at different angles. The analysis of the reflection of a plane wave from a planar interface is also straightforward in the linear case, if the media on each side of the interface are fluids. Using the boundary conditions of continuity of particle velocity and acoustic pressure (see 2.2.2), the pressure amplitude reflection coefficient is given by:

$$\frac{p_r}{p_i} = \frac{Z_2 - Z_1}{Z_2 + Z_1}$$

where p_i is the amplitude of the pressure wave incident on the boundary, p_r is the amplitude of the pressure wave reflected from the boundary and Z_2 and Z_1 are the characteristic acoustic impedances (see 2.2.3) on each side of the boundary (the wave being incident in medium 1).

The transmitted wave amplitude, p_t, is given by:

$$\frac{p_t}{p_i} = \frac{2Z_2}{Z_1 + Z_2}$$

For waves incident at an angle θ_i to the normal to the boundary, there will be a reflected wave (at an angle $\theta_r = \theta_i$) and the transmitted wave will be refracted according to Snell's laws, i.e. it will travel at an angle to the normal given by:

$$\sin\theta_t = \frac{c_2}{c_1}\sin\theta_i$$

where c_1 and c_2 are the speeds of sound in the media on each side of the boundary.

The reflected and transmitted pressure amplitudes are then given by:

$$\frac{p_r}{p_i} = \frac{Z_2\cos\theta_i - Z_1\cos\theta_t}{Z_2\cos\theta_i + Z_1\cos\theta_t}$$

and

$$\frac{p_t}{p_i} = \frac{2\,Z_2\cos\theta_i}{Z_2\cos\theta_i + Z_1\cos\theta_t}$$

If $c_2 > c_1$, total reflection can occur at angles of incidence greater than θ_c when:

$$\sin\theta_c = \frac{c_1}{c_2}$$

If the intensity reflectivity is required, it is simply equal to the square of the pressure amplitude reflectivity:

$$\frac{I_r}{I_i} = \left(\frac{p_r}{p_i}\right)^2$$

The intensity transmissivity of the surface is given by:

$$\frac{I_t}{I_i} = \frac{4\,Z_1Z_2\cos\theta_i\cos\theta_t}{(Z_2\cos\theta_i + Z_1\cos\theta_t)^2}$$

Typical values for the normal incident intensity reflectivity of various materials in water are given in Table 2.3.

If the materials on each side of the boundary are attenuating (see 2.2.5), the above formulae become very much more complicated. The attenuations enter as ratios to the wavelength and introduce extra phase shifts. These appear to be small effects for most tissue boundaries.

If the boundary occurs between a fluid and a solid, the longitudinal wave in the fluid may be partially converted into transverse (shear) waves and

Table 2.3. Intensity reflectivity of different materials in water

Material	Normal intensity reflectivity
Air	1.0
Water	0
Castor oil	0.020
PMMA ("Perspex")	0.37
Aluminium	0.85
Brass	0.93

surface waves in addition to the fraction of the wave that remains as a longitudinal wave. This process is called "mode conversion".

2.2.5 Attenuation

Energy may be lost from a propagating ultrasonic beam either by conversion to other forms of energy in the material (absorption) or by re-direction of small fractions of the beam due to small inhomogeneities in the medium (scattering). The overall effect is such that a plane wave of initial amplitude $A(0)$ will, after travelling a distance x, have an amplitude $A(x)$ given by

$$A(x) = A(0)\exp(-\mu x)$$

This effect is expressed as an attenuation coefficient μ, where

$$\mu = \frac{1}{x}\log_e \frac{A(0)}{A(x)}$$

and is expressed in nepers per metre.

Expressed in decibels per metre, the attenuation coefficient is α, where

$$\alpha = 20(\log_{10}e)\mu = 8.686\mu$$

In homogeneous materials such as fluids, where there is no scattering, it is appropriate to use the term absorption coefficient.

For compressional waves, some typical values for absorption and attenuation coefficients are given in Table 2.4. The loss (absorption or attenuation) in different materials always increases with frequency. For water the absorption coefficient increases as f^2, for castor oil as $f^{\frac{5}{3}}$ and soft tissues approximately as f^1.

As already mentioned in 2.2, shear waves do not readily propagate in soft tissue and are therefore subject to very high attenuation – typically over one thousand times higher than that for compressional waves. Shear waves are of importance in medical ultrasound when compressional waves impinge on bone. Under certain conditions, shear waves will be generated in the bone through

Table 2.4. Typical values of absorption and attenuation coefficients at 1 MHz and 20°C (Np/m)

Material	Attenuation coefficient
Water	0.025
Castor oil	10.9
Blood	1.7
Brain	8.6
Fat	7.0
Heart	23.0
Liver	14.0
Muscle	11.0–16.0
Skull bone	164

"mode conversion". They will then be attenuated in the bone leading to possible heating of the bone and the surrounding tissue.

2.3 Transduction

2.3.1 Piezoelectric Materials

Man's ability to use ultrasonics dates from the discovery by the brothers Curie of the effect of piezoelectricity ("electricity from pressure") in quartz and some natural materials. None of the other methods – mechanical, capacitive, electrodynamic, magnetostrictive – can be used above about 100 kHz. From natural materials we have moved to ceramic materials such as lead zirconate titanate – PZT (in the late 1950s), and on to plastic materials such as polyvinylidene difluoride – pVF_2 or pvdf (in the 1970s). The main characteristic parameters are given in Table 2.5.

The transmitting constant, *d*, is the particle displacement produced in the material per unit applied voltage. For generation, clearly a high value is required although this will also increase exposure. The receiving constant, *g*, needs to be high to be sensitive to small echoes for diagnostic and Doppler devices. In terms of pulse-echo efficiency, the product of *d* and *g* is relevant and PZT is much better than quartz. Pvdf is lower by a factor of almost 3, but has a much higher receiving constant which would tend to indicate a much lower exposure for a given received signal level. In addition to this, the characteristic acoustic impedance of pvdf is much closer to water (or tissue) than that of PZT and the ultrasound is thus much more efficiently transferred between the transducer and the medium being irradiated. The major problem with pvdf arises from its extremely low dielectric constant making it hard to transfer electrical energy into it. The last item in the table is the Curie temperature – the temperature at which the material loses its piezoelectricity. For pvdf there is no such temperature defined. It can still be used up to 130°C but its properties are much less stable above 80°C, hence the brackets.

Table 2.5. Characteristics of piezoelectric materials

	Quartz	PZT4	PZT5A	Pvdf
d (m/V)	2.31×10^{-12}	284×10^{-12}	374×10^{-12}	20×10^{-12}
g (V m/N)	5.78×10^{-2}	2.61×10^{-2}	2.48×10^{-2}	17.4×10^{-2}
k	0.095	0.70	0.705	0.10
ϵ (F/m)	4.0×10^{-11}	1150×10^{-11}	1500×10^{-11}	0.98×10^{-11}
ρ (kg/m^3)	2650	7500	7750	1780
c (m/s)	5740	4000	3780	3000
Z (kg/m^2/s)	15.21×10^6	30×10^6	29.29×10^6	5.34×10^6
T_c (°C)	573	328	365	(80)

2.3.2 Transducer Construction

A transducer made from PZT is constructed (see Fig. 2.2) so that the element will resonate mechanically in a thickness mode at the frequency of interest. This means that its thickness is chosen to be half a wavelength. If continuous waves are required the element has no backing material. After the electrodes are attached, the element is mounted in a holder (case) and a protective layer put over the front surface. This can be chosen to be of a characteristic acoustic impedance between that of the element material and the tissue. If the thickness is properly chosen ($\lambda/4$) the transmission and reception of a particular frequency can be optimised.

In diagnosis, short pulses are needed. Even if the element is given a very sharp electrical stimulus, the element will tend to "ring" at its resonant frequency. This ringing can be shortened electrically or mechanically. The most usual method is mechanical – the mass of the element is simply increased by affixing a backing material to the element. The material has to be highly absorbent to remove the ultrasonic waves emitted by the back of the element and is normally made from epoxy resin loaded with tungsten powder.

The effect of the backing material is not only to shorten the pulse but also to lower the sensitivity of the transducer and often to reduce significantly the central operating frequency. The protective layer (epoxy or plastic) can be used to make very small modifications to the frequency response of the transducer but it can only optimise transmission at one of the broad range of frequencies in a short pulse.

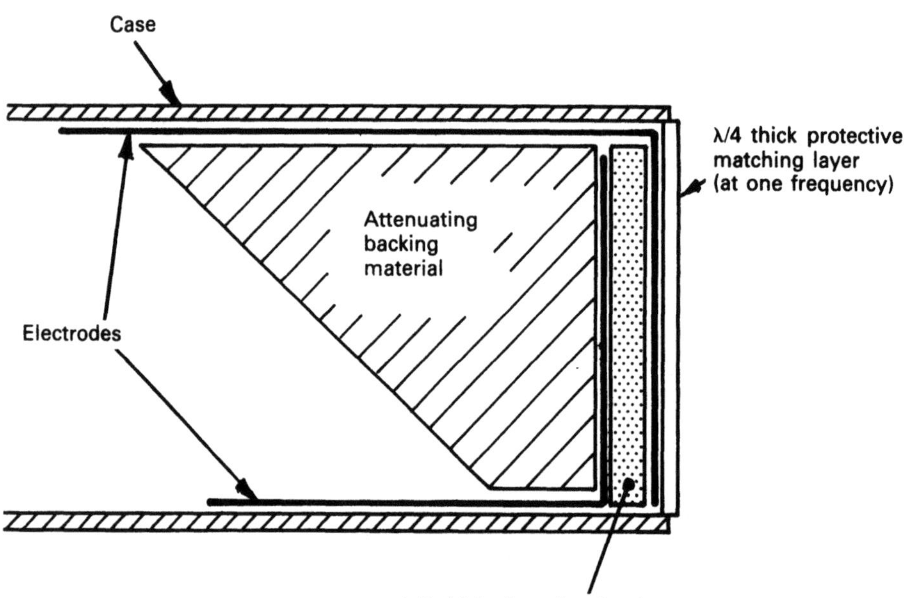

Figure 2.2. Schematic diagram of a single-element transducer construction.

The shape of the elements used is determined by convenience. Single element transducers tend to be circular discs which have a simpler field structure, while arrays are almost always rectangular elements for ease of manufacture.

2.4 Ideal Fields

2.4.1 Sinusoidal Time Excitation

For sinusoidal excitation and referring to Fig. 2.3:

$$p_Q(t) \propto \frac{\rho}{2\pi}\dot{\xi}_0 \int_S \frac{\exp i(\omega t - kr)}{r}\,dS$$

This is known as the Rayleigh integral. $\dot{\xi}_0$ is the normal velocity at the surface which is constant for all points (a piston-like source). In general, this equation cannot be solved in closed form even if the transducer S is a disc. Two cases for the disc can be solved analytically (i) the axial field and (ii) the far field.

2.4.2 Axial Distribution Near Field

For a disc piston source and sinusoidal excitation, the field distribution in the near field is complex and consists of many maxima and minima which can be shown to occur at:

Axial maxima: $$Z_{max} = \frac{4a_1^2 - \lambda^2(2m+1)^2}{4\lambda(2m+1)}$$

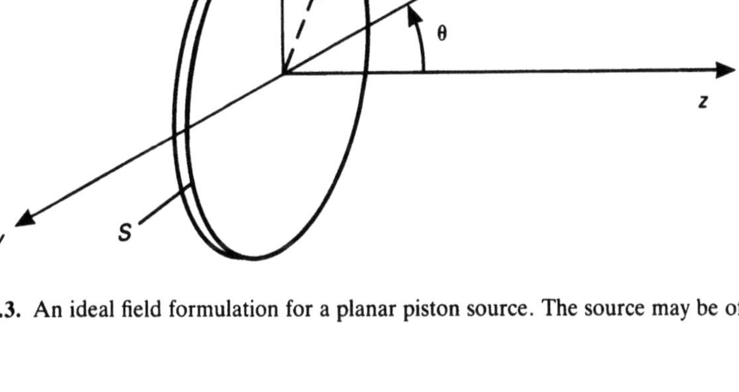

Figure 2.3. An ideal field formulation for a planar piston source. The source may be of any shape.

Axial minima:

$$Z_{\min} = \frac{a_1^2 - \lambda^2 m^2}{2m\lambda}$$

where m is an integer and increases towards the source, a_1 is the transducer radius. These formulae may be obtained by using a direct wave/edge wave model with a π phase shift between the component waves.

The last axial maximum is at

$$Z_0 = \frac{4a_1^2 - \lambda^2}{4\lambda}$$

For $\lambda \ll a_1$, $Z_0 \simeq a_1^2/\lambda$

This distance is usually regarded as the dividing point between the near field and the far field. The far field thus refers to axial distances greater than a_1^2/λ.

In order to maximise the length of the near field, ie. the distance over which the beam is collimated, clearly a_1 should be as large as possible. However, increasing a_1 reduces the lateral resolution one is trying to maintain by the collimation. Some compromise is essential. The axial and lateral pressure amplitude distributions for a disc source are shown in Fig. 2.4.

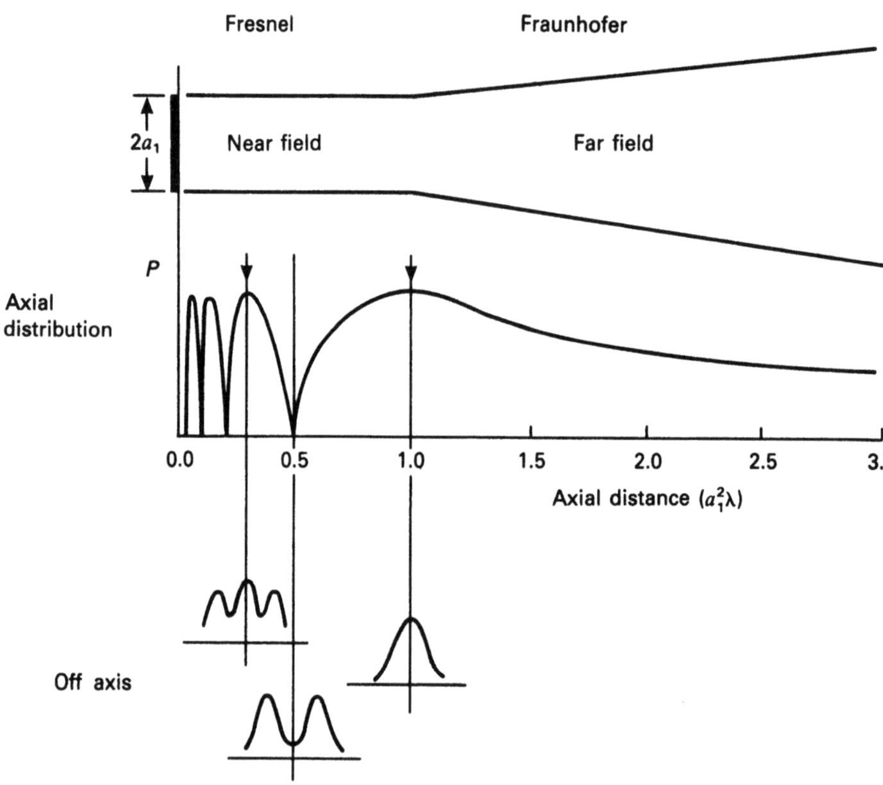

Figure 2.4. Beam patterns and distributions both on-axis and off-axis for a disc piston source.

2.4.3 Far Field

The axial distribution in the far field of a disc can be shown to be of the form:

$$p_Q = \frac{\exp i(\omega t - kR)}{R} \cdot \frac{2J_1(ka_1\sin\theta)}{ka_1\sin\theta}$$

$$= \text{spherical wave} \times \text{directivity function}$$

where J_1 is the 1st order Bessel function of the first kind. The directivity function is shown on a polar diagram in Fig. 2.5.

Whereas the intensity remains reasonably constant in the near field, the existence in the far field of a spherical wave will cause the intensity to decrease with distance from the source. If the effect of the directivity function is small, the decrease will be close to an inverse square law.

2.4.4 Rectangular Sources

For the disc source there is essentially only one dimension of significance – the radius – whose absolute value is only of interest when expressed as a ratio to the wavelength λ (a_1/λ). For a rectangular transducer of width, w, and height, h, there are clearly two independent dimensions of significance. Each of these produces its own near field–far field effects (see Fig. 2.4) and the overall result is an interference of the two. The result is clearly of much greater complexity than for a simple disc.

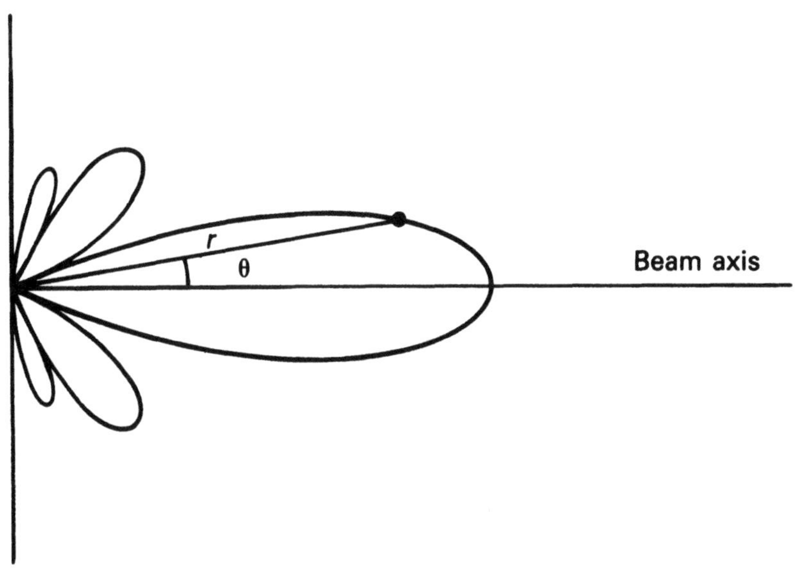

Figure 2.5. Polar plot of the far field distribution from a piston source. The length of the radius vector r gives the (relative) magnitude of the pressure amplitude at a given angle θ.

2.4.5 General Time Excitation

The main problem of the discussion of sections 2.4.1 to 2.4.3 is that it is defined only for a single frequency. A short pulse has a wide range of frequencies and an analysis in the time domain is more appropriate (although equivalent).

For a piston-like source with a surface particle velocity $\dot{\xi}(t)$, the velocity potential at $\underline{r}$ is given by:

$$\phi(\underline{r},t) = \dot{\xi}(t) * h(\underline{r},t)$$

where $*$ represents a convolution, $h(\underline{r},t)$ is an impulse response function and depends on the geometry (of the source and field point Q) only. The pressure amplitude $p(\underline{r},t)$ is given by

$$p(\underline{r},t) = \rho \frac{\partial \phi}{\partial t}(\underline{r},t)$$
$$= \rho \dot{\xi}(t) * \dot{h}(\underline{r},t)$$

In reception, if we have a point source $p(\underline{r},t)$ at $\underline{r}$, the output voltage of a perfect receiver is:

$$V(\underline{r},t) = \dot{h}(\underline{r},t) * p(\underline{r},t)$$

If we have a point reflector (at $\underline{r}$) then:

$$V(\underline{r},t) \propto \underbrace{\rho \dot{\xi}(t) * \dot{h}(\underline{r},t)}_{\text{excitation}} * \dot{h}(\underline{r},t)$$

In the frequency domain we have:

$$\tilde{V}(\underline{r},\omega) \propto \rho \tilde{\dot{\xi}}(\omega)\, \dot{\mathrm{H}}(\underline{r},\omega)^2$$

where $\tilde{}$ represents a Fourier transform and $\mathrm{H} = \tilde{h}$.

Note that these equations assume that there are no losses, and the use of ideal transducers.

2.4.6 Focusing

Focusing can be achieved by shaping the transducer (as a spherical bowl), by using a lens, or in a multielement transducer by changing the relative phases of the waves from the elements in either transmission, reception, or both.

For lenses, which are usually made of plastic or some solid material, a biconcave (or plano-concave) lens is in fact a converging (focusing) lens. This is a consequence of the speed of sound in solids being greater than that in water or tissue (see Table 2.1). Standard formulae from optics can be used to calculate the focal length.

For optically curved elements the field has much the appearance of Fig. 2.4 with a focusing effect superimposed on it. The maximum pressure amplitude (the "true" focus) is usually well inside the geometrical focus (the centre of curvature of the bowl). Its actual position depends upon the ratios of the geometrical dimensions of the bowl to the wavelength.

The focal spot has a finite width. Its shape is actually the "bell" shape of the function:

$$\frac{2J_1\,(ka_1\sin\theta)}{(ka_1\sin\theta)}$$

(where J_1 is a first order Bessel function), and drops to zero at the first zero of the Bessel function. This is usually considered an extreme measure of the focal spot size and the full width at half maximum (*FWHM*) is more often used:

$$FWHM = 0.71\,\lambda\frac{F}{a_1} = 0.71\frac{cF}{fa_1}$$

where λ is the wavelength, F the focal length, f the frequency and a_1 is the radius of the bowl. *FWHM* increases if F increases or a_1 or f decrease.

2.5 Nonlinear Propagation

All the analysis given in 2.2.1 and 2.2.3 is based upon acoustic waves whose amplitudes are small. This "linear" approximation (see Eq. (2.1)) allows us to calculate the effects of a number of waves just by adding them up. (The integral in 2.4.1 is the mathematical formulation of this addition.)

If the waves are not of small amplitude – ie. they are of "finite" amplitude, the simple linear equation is not valid and we have "nonlinear" waves. The second term on the right-hand-side of Eq. (2.1) must then be included. This leads to a much more complicated equation for the waves. When analysed it shows that in a relatively lossless fluid such as water the sine wave which starts from the transducer becomes distorted as it propagates. The compressions travel faster than the rarefactions until it looks more and more like a sawtooth wave. As part of the saw tooth becomes vertical a shock wave forms which is highly disruptive of the medium in which it is propagating.

Another way of describing this phenomenon is to say that as the wave propagates, energy is progressively transferred from the original sine wave to its higher harmonics. Remembering (see 2.2.5) that attenuation in many materials increases rapidly with frequency, the energy in these higher harmonics will be more rapidly attenuated than the energy in the fundamental. If the damping effect of the attenuation is stronger, a true shock-wave will never develop, although very sharp acoustic pressure gradients with potential for damage may still be generated.

2.6 Recommended Additional Reading Material

Cracknell, AP. Ultrasonics. Wykeham Publications Ltd, London (Taylor and Francis Group), 1980

Kinsler LE, Frey AR, Coppens AB, Sanders JV. Fundamentals of acoustics, 2nd edn. J Wiley and Sons, New York, 1982

Wells PNT. Biomedical ultrasonics. Academic Press, London, 1977

3

Overview of Measurement Techniques

B. ZEQIRI

3.1 Introduction

Chapter 2 provided an introduction to the scientific foundation of ultrasonic field measurement by outlining fundamentals concerning the way ultrasound propagates and the way the ultrasound is distributed in a transducer field. This provided a scientific background to the different types of medical ultrasonic systems introduced in Chapter 1 ranging from diagnosis, in obstetrics, to the therapy provided by lithotripters. Whatever the application, of concern to us all is the safe and effective use of ultrasound in medicine. Chapter 1 considered the question of safety of medical diagnostic ultrasonic equipment and the relevance of acoustic output information to the clinical user. The concern over safety and efficiency emphasises both the importance of accurate measurement techniques for determining the acoustic output of equipment and the assessment of which acoustic parameters are most significant for safety. Both topics will be considered in this chapter.

At the present time it is not possible to determine ultrasonic field parameters in tissue or in materials which closely resemble tissue. Consequently, measurements are normally performed in water baths and tissue values sometimes extrapolated using propagation models. This chapter provides an overview of the different techniques which have been used to measure ultrasonic fields. More extensive, and detailed, reviews of the subject together with full lists of references may be found in the literature [1–4]. In this chapter, specific references are given to more recent publications and to work not covered in these reviews.

3.2 Key Acoustical and Other Field Parameters

Before describing the measurement techniques more fully, it would be instructive to consider in rather general terms which properties of the transducer field are currently thought to be of most relevance to safety. Although the key acoustical parameters have been introduced in Chapters 1 and 2, this chapter explains why certain parameters and field properties are important.

Ideally, the full characterisation of the ultrasonic field produced by a transducer would require knowledge of the distribution of acoustic pressure at every point in space and at every instant of time. This is both an enormous task and represents a large amount of data to be stored. In practice, the procedure adopted is to characterise the field in terms of the peak values or spatially- or temporally-averaged values of various acoustic parameters. These will be described in detail in Chapter 6 for typical pulsed fields used in diagnosis. The assessment of any risk to the patient becomes a problem of relating the particular measured acoustic parameter to its possible bio-effect and this will depend on the complex details of the tissue–ultrasound interaction. There are currently thought to be two types of interaction and these are broadly termed thermal and non-thermal effects.

As ultrasound propagates in tissue it is absorbed and this results in a temperature rise dependent on the absorption coefficient of the tissue, incident ultrasonic intensity and the time of exposure. This is a simplified description of what is happening, and other physical properties of the tissue such as thermal conductivity and specific heat capacity clearly play a role. Also important is the way ultrasound is distributed, as indicated by the beam-width, and whether or not the waveform is distorted. The latter is important as the absorption coefficient is very frequency dependent and any high frequency components in the pulse will be readily absorbed and result in excess heating [5]. The acoustical parameters most relevant to the thermal effects are the beam-width and the time-average intensity, and are of greatest importance in physiotherapy. However, the high output levels of some types of Doppler equipment mean that tissue heating generated by these systems cannot be overlooked.

Non-thermal effects are related to the mechanical stress induced in a material during the oscillating displacements caused by the passage of ultrasound. Effects such as cavitation or micro-streaming are considered to be important, although cavitation is, at the moment, thought to be more significant. Cavitation is again a complex phenomenon, depending as it does on the physical state of the tissue and the existence of any bubble nucleation sites as well as acoustic field parameters such as acoustic pressure, with the peak-rarefactive pressure, p_-, thought to be the most significant. Other factors of importance are likely to be the acoustic frequency and the transducer pulsing regime adopted ie. the number of cycles and the repetition period. Clearly, if cavitation is important, it is likely to occur in applications where high peak-pressures are generated, for example in ultrasonic diagnosis and lithotripsy.

There is therefore a need to characterise ultrasonic fields in terms of a range of acoustical parameters for which different measurement methods are employed. The following sections provide an overview of techniques which have been used to characterise the ultrasonic fields in general, although not necessarily from medical transducers. The discussion has been subdivided according to the parameter that the technique measures: acoustic pressure, acoustic displacement, radiation force and temperature rise.

3.3 Acoustic Pressure

3.3.1 Pulse-echo Technique Using a Small Target

The transducer under test may be operated in conventional B-scan mode to detect the echo returning from a fixed target placed in the transducer field. The target, which is normally a sphere or hemisphere, may be moved throughout the transducer field and the echo strength recorded as a function of target position to generate a three-dimensional map of the field distribution. Although the transducer responds fundamentally to the acoustic pressure averaged over its face, the actual field distribution is complex, depending on the transmit and receive properties of the transducer as well as the way the target reflects the ultrasound. Absolute properties of the acoustic field are therefore difficult to derive although the technique may be used as a qualitative check on transducer field patterns. In addition, as the method relies on the detection of the echo by the system under test, it is not possible to apply this technique to certain types of continuous-wave and lithotripter fields.

3.3.2 Piezoelectric Hydrophones

The pressure fields generated by medical ultrasonic equipment may be detected using miniature sensors called hydrophones. These consist of thin plates or films of piezoelectric material fabricated in the form of small sensing elements which convert the incident acoustic pressure waveform into an electrical signal. The simple principle of operation (shown schematically in Fig. 3.1) in combination with the high temporal and spatial resolution makes hydrophones powerful tools in the measurement of all types of ultrasonic fields. Being small and adaptable, they can easily be incorporated in relatively compact

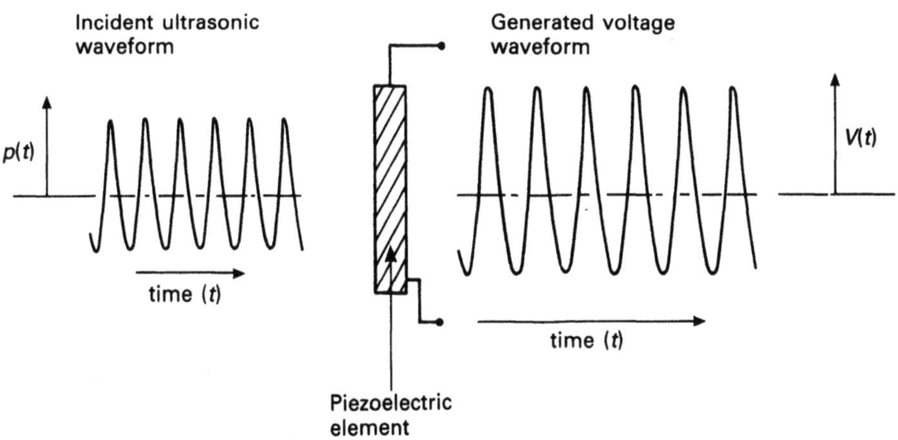

Figure 3.1. Schematic diagram of the piezoelectric element as an input–output device, converting the incident acoustic waveform, $p(t)$, into the observed voltage waveform $V(t)$.

measurement systems (see Chapter 4). Naturally enough, accurate measurements of field properties require that the measured waveform is a faithful reproduction of the incident acoustic waveform.

The extent to which this is achieved depends on the properties of the particular hydrophone being used and it is possible to identify several important characteristics that an "ideal" hydrophone should possess. These are:

1. The device should be sufficiently sensitive to produce a reasonable signal level. The devices are not absolute and their sensitivity, expressed in volts per pascal, must be determined by calibration in a "known" field.
2. The voltage response of the hydrophone should be linear over the range of pressures encountered in medical ultrasonics.
3. The sensing or active element should be smaller than or comparable in size to the acoustic wavelength which, for propagation in water, will vary from 1.5 to 0.1 mm in the frequency range 1 to 15 MHz.
4. The hydrophone should show little variation in sensitivity with frequency over the range 1 to 15 MHz.

Other requirements, perhaps slightly more practical, may also be highlighted:

5. The hydrophones should be robust and be able to operate effectively under the acoustic pressure and power conditions generated by medical ultrasonic equipment.
6. In order to be used as reference working devices their sensitivity should be stable over a significant period of time.

Several designs exist for these hydrophones and their reliability in quantitative acoustic output measurements is now accepted. The choice of hydrophone type depends on the application, as noted in Chapter 4. In practice, none of the different types of hydrophone fully meet the requirements described above; the extent to which they do is described in some detail in Chapter 4.

3.3.3 Optical Techniques

Regions of compression (pressure increase) and rarefaction (pressure decrease) in an ultrasonic wave result in very small changes in the refractive index of the transmitting medium. These provide an opportunity to probe the pressure distribution using optical diffraction, a technique which is particularly attractive due to its non-perturbing nature. Fig. 3.2 shows a schematic of a typical set-up with a parallel beam of light incident on the ultrasound propagating into a tank of water. As the speed of light is very much greater than the speed of ultrasound, the positions of the pressure maxima and minima appear stationary. The ultrasound behaves as a phase grating, introducing phase delays or advances to the optical wavefronts dependent on the local values of the refractive index.

A diffraction pattern is produced at the focus of the second lens with the different orders being detected using a photodetector [6]. It is the intensity of light in these diffraction orders which may be used to derive the acoustic pressure. The theory is complex but for sinusoidal waves at low megahertz frequencies the light intensity in the nth order, I_n, is given by:

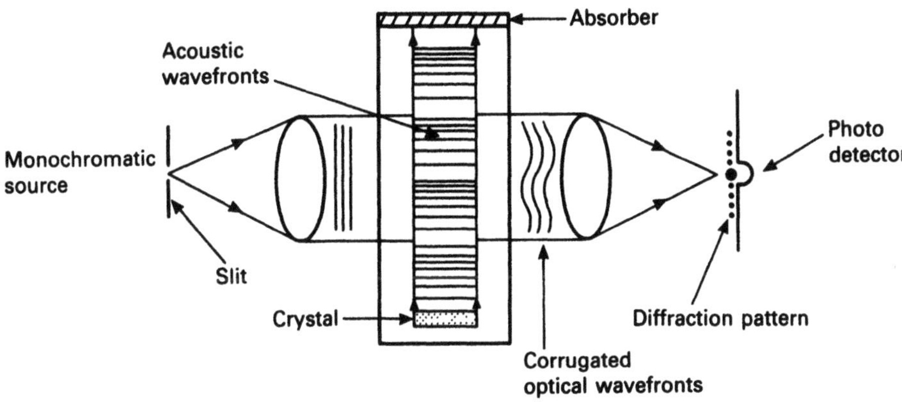

Figure 3.2. Set-up used to determine acoustic pressure using an optical diffraction technique.

$$I_n = J^2{}_n (k\kappa pL) \tag{3.1}$$

where J_n is known as a Bessel function of the nth order, p is the acoustic pressure amplitude, L is the path length, known as the interaction length, κ is a parameter known as the adiabatic piezo-optic coefficient and k is the optical wavevector. Eq. (3.1) illustrates one of the main shortcomings of the technique: the determination of the parameter L. It represents a line-of-sight integration over the beam cross-section and its estimation usually makes assumptions about the beam shape. Additional limitations include an inability to resolve pulsed waveforms whose length is less than the optical beam-width and a difficulty in application to focused fields.

Another optical technique which may be used to characterise transducer fields is the schlieren technique and this uses an experimental set-up very similar to Fig. 3.2. The difference is that the non-diffracted part of the optical beam is blocked by a "stop" so that only light which has been diffracted by the ultrasound reaches the screen providing a visualisation of the ultrasound. The technique is qualitative but provides information, useful for teaching purposes, of the way ultrasound propagates both in pulsed and continuous-wave excitation. It may also be used to detect gross imperfections in the beam profile. Although schlieren systems are traditionally large, it is possible to produce small portable units which are particularly suited to the monitoring of the ultrasonic fields generated by physiotherapy systems [7].

One of the attractions of any technique which utilises the interaction of optical and acoustic wavefronts is clearly its non-invasive nature. This raises the possibility of characterising acoustically hostile environments where conventional measuring devices may suffer damage. An example of such an environment is the focus of a shock-wave lithotripter where peak-compressional pressures of 100 MPa may be generated which may shorten the working lifetime or even destroy valuable reference devices such as hydrophones.

It has only been possible to describe a few of the large number of optical techniques which have been applied to the measurement of ultrasonic fields. However, a recent development worthy of mention is the fibre-optic hydrophone

[8] which utilises the optical birefringence induced in a thin optical fibre by ultrasound to record the pressure in the field.

3.4 Acoustic Displacement

3.4.1 Optical Interferometry

A thin acoustically transparent membrane or pellicle, when placed in an acoustic field, will move in unison with any movement of the surrounding medium and will therefore follow the particle displacement as the ultrasound propagates. For continuous plane progressive waves of intensity 100 mW/cm^2 the resultant displacement amplitude at 5 MHz is approximately 1 nm and may therefore be measured using optical interferometry. Fig. 3.3 illustrates the principle behind the technique, with the displacement being calculated from the measurement of the phase difference between a fixed optical reference beam and an optical beam which has been reflected from the surface of a metallised moving membrane or pellicle.

Assuming plane progressive waves, the acoustic pressure is derived from the displacement using the relationship:

$$p = \rho c \omega d \tag{3.2}$$

where d is the measured amplitude of the particle displacement, p is the acoustic pressure amplitude, ρ is the density of the propagating medium and ω is the angular frequency. As well as the direct and absolute measurement of particle displacement, the technique offers high resolution, both spatial (0.1 mm) and temporal (10^{-8} s). Unfortunately, the systems are, in practice, complex and expensive, partly due to the stringent need for isolation from environmental vibration. Their use is therefore likely to be restricted to specialised laboratories. At NPL, a specially-designed laser interferometer is used as a primary standard instrument for calibrating hydrophones [9].

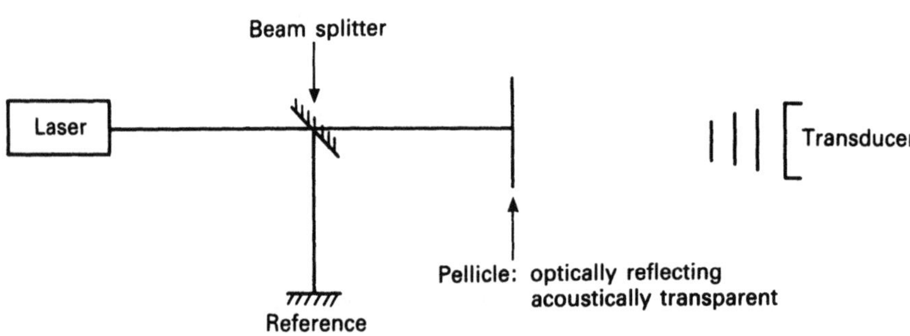

Figure 3.3. Diagram showing an optical interferometric technique for measuring the acoustic displacement of a thin membrane placed in a transducer field.

3.4.2 The Capacitance Probe

The acoustic displacements produced by ultrasound in the megahertz frequency range may be measured using a capacitance probe. The device is essentially a parallel plate capacitor with a small, normally air-filled, separation of typically 0.1 mm. A typical polarisation voltage of 100 V is applied across the plates such that when ultrasound is incident on one plate the gap-width is modulated and a voltage change measured. Advantages of the technique are its absolute nature, its ability to detect over a wide frequency range and its high sensitivity although the latter is unfortunately only achieved by detecting over a large area (typical active elements are of 5 mm diameter). The high spatial resolution required for characterising some medical ultrasonic systems would mean using very small elements which in turn would make the probe difficult to use. Additionally, its use for measurements in water requires the ultrasound to be coupled to the probe by a buffer block whose acoustic transmission properties must be determined. To date, the main applications of such a device have been in nondestructive testing, for example the detection of acoustic emission and laser generated ultrasound. However, due to their ruggedness and durability, capacitance hydrophones have recently been used to investigate lithotripter fields [10].

3.5 Radiation Force

The propagation of ultrasound through a medium is associated with a simultaneous transfer of momentum. Momentum is transferred to any absorbing or reflecting obstacle which intercepts the ultrasonic beam and this is revealed as a radiation force acting on the obstacle. The measurement of this force provides the basis of a number of acoustic intensity or power measuring systems. The force actually depends on the time-averaged intensity in the ultrasonic beam. Two basic types of measurement system may be defined dependent on how much of the ultrasonic beam is intercepted by the obstacle or target:

1. *Large-target*: The target intercepts the whole of the ultrasonic beam to yield a value of the total acoustic power.
2. *Small-target*: The target is much smaller than the spatial extent of the beam and yields the time-averaged values of the local intensity.

These will now be described along with a third technique which uses the deformation of an air–water interface produced by ultrasound striking it as the basis of measuring the transducer beam profile.

3.5.1 Large Target – The Radiation Force Balance

The principle behind the force balance is the measurement of the force experienced by the target which totally intercepts the ultrasonic beam. The force, F, is related to the acoustic power, W, by the expression:

$$F = \frac{hW}{c} \tag{3.3}$$

where c is the speed of ultrasound in the propagating medium, usually water. The quantity h depends on the geometry of the target and is equal to unity for a target which is either reflecting at 45° (plane or conical) or totally absorbing. The possible values of h and how they depend on the geometry of the target are considered in some detail in Chapter 5 along with a description of the design and use of several radiation force balances. To give some idea of the magnitude of this force, a total power of 1 mW will produce a 0.69 μN force on an absorbing target. Many radiation force balances have been described in the literature, differing in design features such as the method of measuring the small force, the geometry of the target and the portability of the measuring instrument. They represent an important class of measuring instrument and in general provide a convenient method for measuring the total output power of a transducer on a routine basis.

3.5.2 Small Target – The Suspended Sphere Radiometer

The force experienced by a small target may be used to determine local values of the acoustic intensity. The target, which should ideally be small in comparison to the spatial variations of pressure in the field, is normally a few wavelengths in diameter. A typical configuration used is shown in Fig. 3.4, and shows a

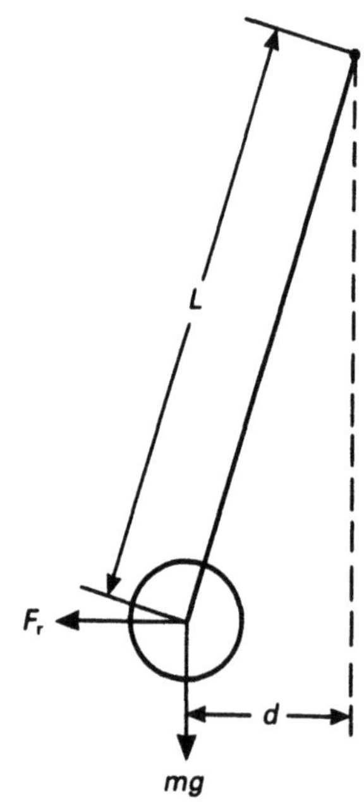

Figure 3.4. The suspended sphere radiometer showing the forces acting upon a sphere suspended in a sound field.

sphere suspended pendulum-fashion from a thin filament, either by direct cementing to the target or by a small hole drilled in the target [11].

The resultant horizontal deflection, d, is measured and used to derive the radiation force F_r given by:

$$F_r = \frac{mgd}{(L^2-d^2)^{\frac{1}{2}}} \tag{3.4}$$

In theory, any shape of target may be used, although to relate F_r to the local value of intensity the interaction between the acoustic field and the target must be amenable to theoretical analysis. Consequently, simple spherical targets are normally used and calculations exist for a range of materials [12,13]. The calculations are used to derive a quantity, Y_p, known as the acoustic radiation force function, which is given by:

$$Y_p = \frac{F_r}{\pi a^2 \bar{E}} \tag{3.5}$$

and is the radiation force acting on the sphere per unit cross-section area per unit mean energy density, $\bar{E}$, of the field of interest. The time-average intensity in a plane progressive wave, I, is related to $\bar{E}$ by $I = \bar{E}c$ and hence I is given by:

$$I = \frac{F_r c}{\pi a^2 Y_p} \tag{3.6}$$

The reliability of the calculations will depend on how closely the experimental configuration mimics theory. In this respect, particular attention must be paid to the suspending filaments to ensure that they are thin enough to be non-perturbing and that the method of attachment produces no departures from target sphericity. Calculations of Y_p, plotted against the quantity ka, are shown in Fig. 3.5 and illustrate the complex nature of its frequency dependence. The positions of the minima are particularly sensitive to the acoustic properties of the target material. Indeed, the influence of the type of material on Y_p has been the subject of extensive study with [14] suggesting materials for which the minima would be shallower.

The technique has relatively low sensitivity, which means that it cannot be used for acoustic output measurements of medical diagnostic ultrasonic equipment, although it may be applicable to measurements on certain types of high power ultrasonic fields such as those generated by hyperthermia devices. Typically, however, the deflections are small and difficult to measure. For a stainless-steel target of radius 0.8 mm, taking a value of L of 110 cm and intensity of 100 mW/cm^2 at 5 MHz the predicted deflection is approximately 0.9 mm. For diagnostic equipment, where the ultrasound is applied in short duration pulses, time-averaged intensities at the spatial-peak may be as low as 1 to 10 mW/cm^2 and corresponding target deflections would be very small. Additionally, the response of the target to the pulse would be a complex function of frequency: a result of the frequency dependence of Y_p (see Fig. 3.5) and the many frequency components normally present in the pulse. For these reasons, use of the small target radiometer has been restricted to the measurement of high power continuous-wave fields or to the calibration of hydrophones where, after the intensity has been measured at the particular

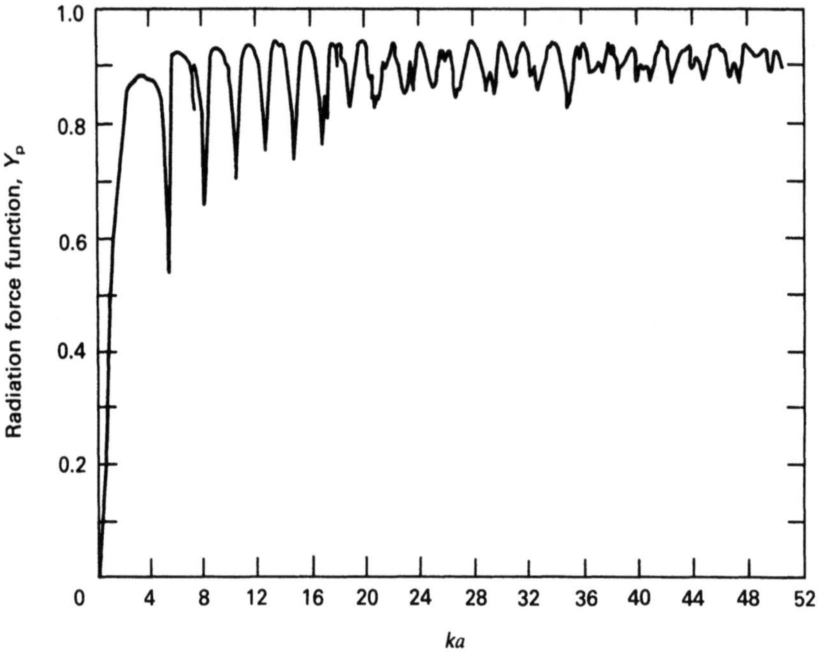

Figure 3.5. Diagram illustrating the complex frequency dependence of the radiation force function, Y_p, for 440C stainless steel as given by Dunn et al. [11]. Y_p has been plotted as a function of ka where k is the wavenumber and a is the radius of the sphere.

point in the acoustic field, the radiometer is replaced with a hydrophone whose output voltage is measured, thereby providing a calibration of the hydrophone.

3.5.3 Surface Levitation

A technique which has been used for quantitative measurements of acoustic fields is surface levitation and involves the ultrasound being directed at a water–air interface. The acoustic radiation force distorts the interface producing a relief map whose profile is a measure of the intensity distribution in the beam. The exact height of any point on the raised water profile depends on a balance of hydrostatic pressure, radiation force and surface tension. For typical powers in the medical range, the displacements are of the order of 10 to 20 μm in the centre of the beam and may therefore be readily probed using optical techniques. An elegant technique which has been applied to the determination of the surface relief profile is called double pulsed holography and involves photographing the flat and raised water surfaces and superimposing the two images during reconstruction. The resulting interference pattern provides a contour map of the beam profile, the displacements at each point being related to the local intensity [15]. However, the technique requires a large and sophisticated facility and is not appropriate for routine use.

3.6 Temperature Rise

As ultrasound propagates in a medium its amplitude decays due to attenuation or absorption in the medium. In a homogeneous medium, where scattering is negligible, the absorption may be due to viscosity, thermal conductivity or molecular relaxation processes. This transfer of energy results in a temperature rise whose measurement forms the basis of two classes of "thermal" measurement technique.

3.6.1 Calorimetry

Calorimetry provides a conceptually attractive way of measuring the acoustic power produced by transducers. In theory, it is necessary only to ensure that all of the acoustic energy is converted to thermal energy by absorption in an absorbing liquid such as castor oil or carbon tetrachloride (solids may also be used). The power may then be derived from the resultant increase in temperature. In practice, the thermal losses are difficult to control and the temperature rise may be difficult to measure accurately, particularly for diagnostic equipment where powers may be as low as a few milliwatts. However, the latter will not be a limitation for measurements on Doppler devices, where acoustic powers are generally greater than 30 mW, or for physiotherapy devices.

Two designs of calorimeter may be identified.

1. The liquid in the absorbing reservoir is stirred during irradiation of the ultrasound and the final mean temperature rise determined. Problems arise due to the difficulty in stirring viscous liquids and the slow response time of the device.
2. The most favoured set-up is the continuous-flow calorimeter, sketched schematically in Fig. 3.6. Liquid passes through an absorption chamber at a fixed rate and the input and output temperature are monitored. Both sophisticated systems [16] and small potentially portable systems have been built.

The devices are usually calibrated using an electrical heater supplying a precisely known amount of electrical energy to produce the same average temperature rise. The technique measures only total power and consequently gives no spatial or temporal resolution.

3.6.2 Thermal Probes

Local measurements of acoustic intensity may be made using small thermoelectric probes embedded in suitable absorbing media. In an early design of Fry and Fry [17] shown in Fig. 3.7, a thermocouple is in castor oil between two polyethylene diaphragms, used to minimise reflection losses. Fig. 3.8 shows how the thermoelectric voltage generated by a one second burst of ultrasound varies with time. Two distinct regions may be defined: the initial rise which

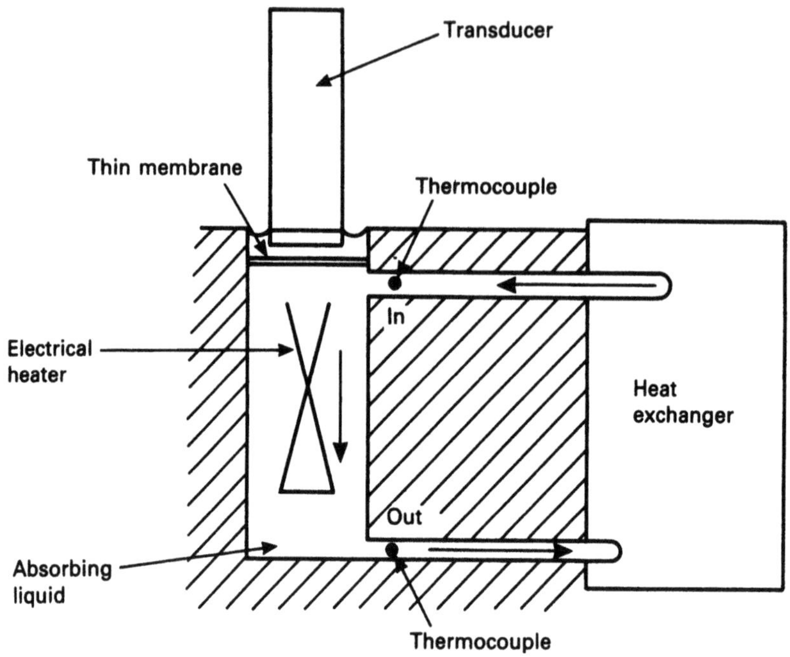

Figure 3.6. Schematic diagram showing the principle of operation of the continuous-flow calorimeter. The total acoustic power is derived from the difference in temperature of the liquid flowing into the calorimeter to that flowing out.

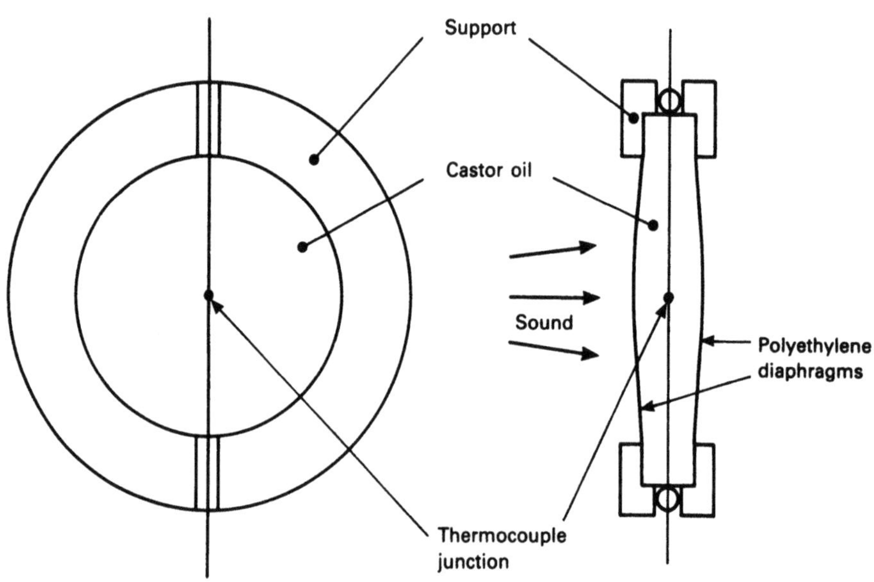

Figure 3.7. Set-up used for thermal-probe measurements showing the thermocouple junction immersed in castor oil, a highly absorbing liquid, enclosed between two polyethylene acoustic windows [17].

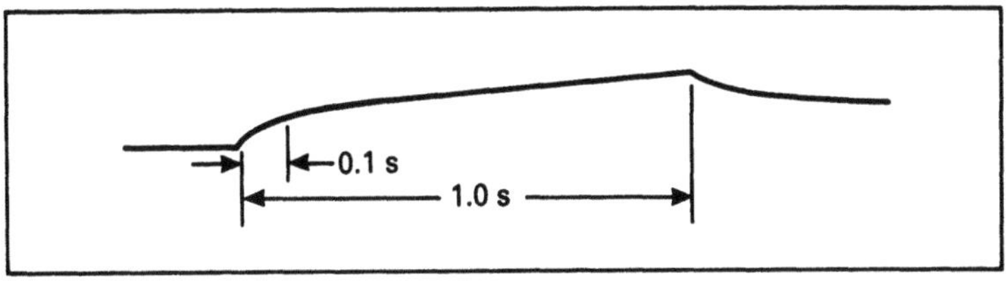

Figure 3.8. Typical thermocouple trace showing the large initial temperature rise followed by a region during which temperature increases linearly with time. The latter is used to determine the local value of the time-averaged intensity.

has been attributed to a viscous interaction between the medium and the wire and a long, almost linear, region, where the temperature rise is due to absorption of ultrasound in the surrounding medium. The slope, that is the rate of temperature rise, of the latter region gives a measure of the local time-averaged acoustic intensity.

The attraction of such a technique is its simplicity coupled with the fact that the thermocouples themselves may be made as small as 10 μm thus enabling very good spatial resolution to be obtained. However, Fig. 3.8 demonstrates one of the main disadvantages, the slow response time which means that the time variation of pulses cannot be resolved. Furthermore, the very fine thermocouples are not easy to manufacture and suffer problems with reproducibility of performance. In principle, for a liquid whose absorption coefficient is known, the devices could produce absolute measurements of intensity, but in practice they are calibrated using a small sphere radiometer. The low sensitivity of these devices is an added problem; it may be overcome by using a thermistor coated in a bead of absorbing medium but the increase in sensitivity is only achieved with a loss in spatial resolution (the bead diameter is of the order of 0.7 mm).

Over the last few years the time-averaged intensities generated by modern Doppler devices have increased to such an extent that there are now concerns about heating effects due to absorption of the ultrasound. It is therefore likely that this type of measurement technique will increase in importance. By embedding the measuring thermocouple in a tissue-like medium which may itself form part of a phantom representing the tissue path to the particular site of interest, it will be possible to measure the temperature rise directly.

3.7 Summary of Quantitative Measurement Techniques

An attempt has been made to review the various measurement techniques available for the characterisation of ultrasonic fields. It is sometimes difficult to decide which techniques are most appropriate for a particular application, and to provide guidance Table 3.1 summarises the techniques considered in this chapter. The four major columns follow the previously adopted format in

Table 3.1. Summary of key acoustical and other field parameters, their methods of measurement, and the acoustical parameters which may be derived from them. The directly measured quantities are indicated, along with those that may be derived using the assumption of plane-waves and by scanning the measuring device through the acoustic field. The column headings also give the reference to the sections of the text dealing with the particular measurement techniques

Derived parameter	Parameter measured and method of measurement						
	Pressure		Displacement	Radiation force		Temperature rise	
	Hydrophone 3.3.2	Optical diffraction 3.3.3	Optical interferometry 3.4.1	Small target 3.5.2	Large target 3.5.1	Thermal probe 3.6.2	Calorimetry 3.6.1
Pressure	Direct	Direct but a line-of-sight integration	Assume plane-waves	Assume plane-waves		Assume plane-waves	
Displacement	Assume plane-waves		Direct	Assume plane-waves		Assume plane-waves	
Intensity	Assume plane-waves		Assume plane-waves	Direct time-averaged intensity		Direct time-averaged intensity	
Total power	Assume plane-waves and integrate over plane		Assume plane-waves and integrate over plane	Integrate over plane	Direct	Integrate over plane	Direct
Beam-width	Beam-plot		Beam-plot	Beam-plot		Beam-plot	

describing the class of technique in terms of the parameter measured: pressure, displacement, radiation force or temperature rise. These are further divided into the specific techniques considered, along with the relevant section in the text. Each row refers to one of the parameters required for ultrasonic field characterisation. The purpose of Table 3.1 is to indicate which of the particular measurement methods allows the direct or indirect determination of the parameters of interest. Other chapters in this book are concerned with the use of hydrophones and power balances for measuring acoustic output. For hydrophone measurements, the acoustic pressure may be determined at a position in the transducer field and the intensity at that point derived assuming plane progressive waves (see 2.2.3).

By moving the hydrophone through the transducer field and integrating the intensity over the beam the total power of the transducer may be derived. In fact the technique, which is known as planar-scanning, may be used to calibrate a hydrophone using a transducer of known output power (see 4.3.8). However, the planar-scanning technique is time-consuming and for routine measurement of acoustic power output, the power balance is more convenient and easier to use (see Chapter 5).

Apart from the scientific and technical merits of the different measurement techniques, as outlined in Table 3.1, there are other considerations which may determine which technique is most suitable for a particular application. An attempt has been made to assess these in Table 3.2 which gives a breakdown in terms of cost and technical considerations. The former may be further subdivided into the capital costs of buying the item of equipment and additional investment costs which must be met before the device can be used for measurements. A hydrophone, for example, needs to be held in a special mount that will provide the necessary degrees of freedom of movement (see Chapter 4); the ability to position and move it accurately in the transducer field must be provided, and the voltage waveform generated by the hydrophone in the transducer field will need to be analysed to derive the acoustic parameters

Table 3.2. An assessment of some of the measurement techniques described in this chapter scored against various cost and technical considerations. The more asterisks appearing in any box, the better. The table is described more fully in the text

Consideration		Piezoelectric		Radiation force		Temperature rise	
		Hydrophone	Ultrasound beam calibrator	Large-target	Small-target	Thermal probe	Calorimeter
Cost	Low capital cost	***	*	***	****	***	***
	Low investment cost	**	****	***	***	**	****
Technical	Wide applicability	****	****	**	*	**	*
	Fidelity	***	***	**	*	*	**
	Ease of use	**	****	***	**	**	**

described in Chapter 6. These additional costs must be met before the hydrophone can be used effectively for acoustic output measurements.

The technical considerations have been broken down into the following categories:

1. *Range of applicability*: How versatile is the piece of measuring equipment? Can it cope with the range of medical ultrasonic systems encountered?
2. *Measurement fidelity*: How accurate is the measurement? Is the technique simple in concept or are there a number of sources of measurement uncertainty which will significantly degrade accuracy? If there are sources of measurement uncertainty, can these be reliably quantified and the overall accuracy of the technique assessed?
3. *Ease of use*: Is the technique easy to use or does it require a skilled operator which would increase the overall investment cost?

For hydrophones, after buying the device (and normally a signal amplifier), the additional investment costs are high and involve setting up a signal acquisition and analysis system as well as a stepper motor facility for beam-plotting. Table 3.2 is meant only as a rough guide to the attributes of each of the techniques but even so it is clear the final choice of technique will be a compromise of many factors.

Qualities which have been omitted from Table 3.2 are the speed with which measurements may be made and the portability of the measuring device. The need to scan a single-element hydrophone through the transducer field to determine the acoustic output makes the technique generally time-consuming. An alternative approach is a hydrophone consisting of a linear array of active elements which enable the pressure distribution to be sampled instantaneously across the beam profile. The system referred to as the NPL Ultrasound Beam Calibrator in Table 3.2 uses such an approach. The outputs from the multi-element hydrophone are sampled using fast data capture and are rapidly analysed to give all of the relevant acoustical parameters in a matter of seconds. The system is described briefly in Chapter 4. Although the hardware associated with the NPL Ultrasound Beam Calibrator is significant, it is a transportable system and may be used in a mobile service visiting hospital departments, thereby permitting on-site scanner characterisation to be carried out. This is an important requirement as medical ultrasonic scanners can be in almost constant use and it is clearly inconvenient for the scanner to be removed from the clinical environment for any extensive length of time. Similarly, measuring instruments which are compact and may be kept close to the ultrasonic device for day-to-day routine performance measurements are attractive. The small power balances which will be described in Chapter 5 fall into this category.

3.8 Assessment of Measurement Uncertainty

One of the technical considerations presented in Table 3.2 is measurement technique accuracy. The assessment of measurement uncertainty represents a very important, and often neglected, part of the measurement process.

Whenever measurements are undertaken, and objective decisions are based on the results, a statement of measurement uncertainty is vital. As an example, one important area is compliance testing of ultrasonic equipment, measurements made to ensure that the output levels of a particular item of equipment are in accordance with declared values or are below a specified level. A regulatory system currently operating in the USA is the 510(k) premarket notification required by the Food and Drug Administration [18]. Accuracy is therefore important when comparing measured values with declared values. There are other areas in which it is possible to identify a need for accurate measurements, two of these are the objective assessment of bio-effect experiments and improved definition of "dose" received by a patient. As the theme of accuracy and estimation of uncertainties will be emphasised throughout the book, a brief introduction will be given here and more details are given in Chapter 7.

The overall measurement uncertainty may be broken down into two separate components, referred to as the random and systematic uncertainties. The random uncertainty is a measure of the repeatability of the measurement and is assessed by performing several repeats of the measurements and analysing the data statistically. Systematic uncertainty may originate from a number of different sources dependent on the measurement technique itself and, as the name implies, results in a systematic bias in the measurement. Sources of systematic uncertainty for each technique must be identified and assessed, either by separate measurements or theoretically. A useful treatment of measurement uncertainties and conventions for combining them to give an overall uncertainty may be found in the National Measurement Accreditation Service guidance document B3003 [19].

From the range of techniques considered in this chapter it will be evident that measurements of the acoustic output of medical ultrasonic equipment have been traditionally difficult to make and because of this overall measurement accuracies have been poor. It is only relatively recently with the development of reliable measuring hydrophones that measurement procedures have become better defined and reliable measurements possible. Hydrophones were introduced very briefly in 3.3.2 and are covered in greater detail in Chapter 4.

3.9 Measurement Traceability

To ensure that comparisons between measurements made by different laboratories or manufacturers are valid, there is a need for common measurement standards and hence the concept of traceable measurements is important.

In the UK, measurement traceability from national measurement standards to the end user is provided by the National Measurement System (NMS). At the apex of the NMS is the National Physical Laboratory (NPL) which maintains the majority of national standards for important physical and engineering quantities. NPL disseminates its standards either directly to the end user via its calibration services, or indirectly via the commercial calibration laboratories that use NPL standards. The NMS provides a consistent basis of measurement within the UK but part of the remit of NPL is the coordination of the UK system with the measurement systems of other countries.

For acoustic output measurements made in the UK using hydrophones, traceability should ultimately be to the national standards for hydrophone calibration held at NPL and Fig. 3.9 shows the traceability chain from NPL to the end user (customer). Traceability is to the "primary standard" measurement technique – the NPL laser interferometer briefly mentioned in

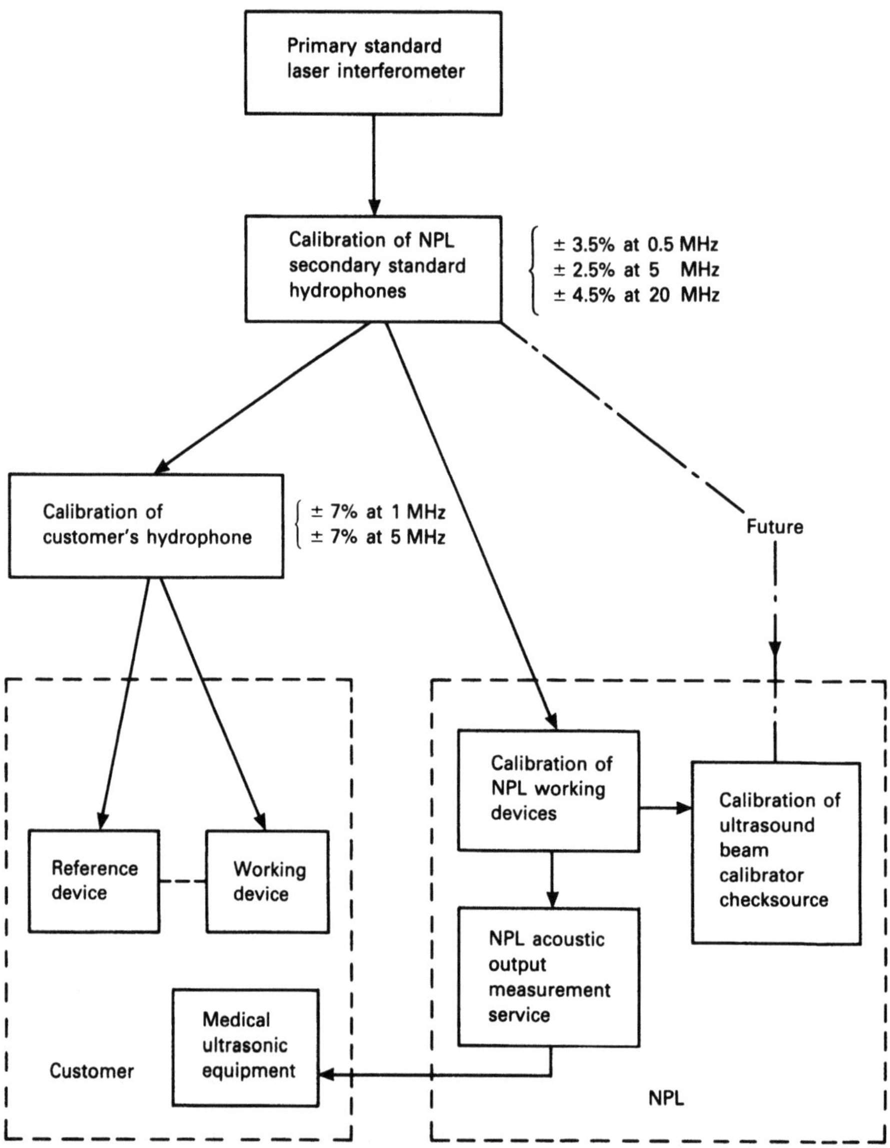

Figure 3.9. Traceability chain for acoustic field measurements made using hydrophones.

3.4.1 and 4.3.8. The interferometer is used to calibrate a few "secondary standard" hydrophones at NPL to an accuracy of between 2.5% and 4.5%. It is these "secondary standard" hydrophones that are used to calibrate both customers' hydrophones, allowing dissemination of the standard, and the NPL "working" standard devices which are used to provide an acoustic output measurement service for ultrasonic equipment. The calibration is actually performed using a substitution method whereby the "secondary standard" and customer hydrophone are placed sequentially at the same position in an acoustic field and their output voltages compared. Fig. 3.9 illustrates the hierarchical nature of the calibration chain with the most accurate, "primary standard", technique at the top and the accuracy of the hydrophone calibration gradually degrading as we move down the calibration chain. Currently, both the NPL Ultrasound Beam Calibrator (BECA) and the NPL ultrasound checksource, a transducer and electrical drive unit whose output power is stable, are calibrated against one of the NPL "working" standard devices although ultimately improved calibration accuracy will be achievable by calibration against a "secondary standard" hydrophone.

Customers may also have their own hierarchical system in place, which may consist of a single reference standard device and several "working" standard devices that they actually use for acoustic output measurements. The "working" standard and reference standard devices may be compared periodically to check for consistent performance although it is important that both are recalibrated at suitable intervals, preferably once a year.

The concept of traceability embodied in Fig. 3.9 and its application to measurements in medical ultrasound is a fairly new one: calibrations of the membrane hydrophones described in Chapter 4 were first performed at NPL in 1980 and measurements have only been traceable to the NPL "primary standard" interferometer since 1985. Over these years, there has been a gradual build-up of traceability so that, today, the majority of ultrasonic field measurements undertaken in the UK use NPL-calibrated hydrophones. Indeed, a large number of measurements outside the UK are also now traceable to NPL. However, the relatively young state of development of the calibration chain shown in Fig. 3.9 is reflected in the fact that currently hydrophones are still calibrated directly at NPL. Eventually, with the proliferation of hydrophone usage, independent laboratories may seek formal accreditation through the National Measurement Accreditation Service (NAMAS) for performing hydrophone calibrations ultimately traceable to NPL.

3.10 Qualitative Techniques

As well as the various quantitative measurement techniques which can be applied to the characterisation of ultrasonic fields, much simpler methods may be used to obtain more limited, qualitative, information about the way an ultrasonic transducer is operating. The decision concerning whether one applies these simpler methods or utilises any of the quantitative techniques considered earlier in this chapter depends on the nature and the extent of the information required.

At this stage it may be useful to consider the example of a physiotherapy system operating in a hospital department and the hierarchy of measurements which may be carried out to assess its performance and so give the physiotherapist information on patient exposure. The seemingly most trivial test would be to ensure that the physiotherapy treatment head is emitting ultrasound at all and this may be done by observing the cavitation bubbles and streaming set up in a beaker of tap water or in a layer of water placed on the transducer surface. A simple test, maybe, but clearly vital when assessing whether or not the ultrasound therapy unit was functioning both before and after the patient underwent a prescribed treatment.

Another aspect of interest could be the determination of just how ultrasound is distributed as it emerges from the transducer face, ie. is the ultrasound confined to a cylindrical volume perpendicular to the surface of the treatment head? Or, due to manufacturing inadequacies or a cracked crystal, is the beam skew? Are we getting "accidental" focusing which can result in "hot-spots" raising the possibility of local tissue damage? This type of information is useful to the physiotherapist who can then have greater confidence in the treatment. For this purpose, a hydrophone, even an uncalibrated one, can be used to scan the ultrasonic field and provide useful information about the transducer beam pattern and, if scanned systematically over the beam, can provide an estimate of the cross-sectional area of the beam.

The next tier of information may be quantitative information concerning the output power of the transducer derived from measurements using one of the radiation force balances considered in Chapter 5. These measurements may be made routinely to compare long-term device performance or, on a newly purchased device, to ascertain the accuracy of the dial settings of a physiotherapy system and therefore to test the technical specification.

Finally, the acoustic pressures generated by the transducer may be determined using a calibrated hydrophone. Typically, measurements of the peak pressures and intensities found in the field may be determined. Clearly, quantitative measurements of acoustic power and pressure are vital when any assessment of treatment safety is made. However, they are not easy or quick to undertake.

Qualitative methods can play an important role in performance assessment, particularly so when one considers that they are frequently simple tests and can be performed inexpensively and quickly. The lack of reliance on expensive and complex instrumentation is one of the attractions of a technique suggested by Sarvazyan et al. [20] which provides rapid visualisation of the ultrasonic field pattern. The method operates on the principle that ultrasound can accelerate the rate of absorption of water by a porous medium, the rate being proportional to the local time-averaged intensity. If the water is coloured by a dye (Sarvazyan has used a 0.1% solution of methylene blue) the intensity maxima and minima will be distinguishable. By suspending the porous sample (paper or cloth) perpendicular to the acoustic axis, a cross-section of the ultrasonic field at any point is obtained. Although used in this way the technique is qualitative, Sarvazyan has suggested that it may be calibrated against another absolute technique for determining local intensity or the intensity derived from pressure measurements. Exposure times vary with the values of the time-averaged intensities but the technique is probably most useful for physiotherapy fields. In fact, the Sarvazyan technique is an adaptation of the "bleach-out" and "starch-plate" methods that have been used to visualise acoustic diffraction patterns. A short description of the techniques may be found in Carlin [21].

Some materials change colour when subjected to ultrasound, a phenomenon which has at times in the past been referred to as chromotropism. The colour change is due to the temperature rise produced by the absorption of ultrasound and can be used as an indicator of ultrasonic intensity. This is the principle behind the Transducer Strain Gauge manufactured by Ohmic Instruments.* The material used in this device is a liquid crystal and displays colour changes from red to green to blue as the incident acoustic intensity increases from low through medium to high. It can be used to give an approximate value of the area over which the ultrasound is distributed and the acoustic power.

The emphasis throughout this chapter has been upon the quantitative measurement of ultrasonic fields and because of this we have only touched upon a few qualitative techniques for investigating ultrasonic fields. In general, only techniques based on the physical effects of ultrasound have been considered here. However, it is worth pointing out that ultrasound can induce a plethora of effects in different materials, many of which are chemical in nature. These include: speeding up of chemical reactions; separation of chemicals; modification of electrochemical processes; emulsification; luminescence and coagulation. In the past, some of these have been used as the basis of measurement techniques. The interested reader is referred to older text-books such as Bergmann [22] and Carlin [21] for a description of these phenomena.

3.11 References

1. O'Brien WD. Ultrasonic dosimetry, In: Fry FJ (ed) Ultrasound: its applications in medicine and biology, part I. Elsevier, Amsterdam, 1978, pp 343–391
2. Stewart HF. Ultrasonic measurement techniques and equipment output levels, In: Repacholi MH, Benwell DA (eds) Essentials of medical ultrasound. Humana, New Jersey, 1982, pp 77–111
3. Hill CR. Detection and measurement of acoustic fields, In: Hill CR (ed) Physical principles of medical ultrasonics. Ellis-Horwood, Chichester, UK, 1986
4. Preston RC. Measurement and characterisation of the acoustic output of medical ultrasonic equipment, parts I and II. Med Biol Eng, (Part I) 1986; 24:113–120; (Part II) 1986; 24:225–234
5. Bacon DR, Carstensen EL. Increased heating by diagnostic ultrasound due to nonlinear propagation. J Acoust Soc Am 1990; 88:26–34
6. Reibold R, Kwick P. Optical near-field investigation into the Raman-Nath and KML regimes of diffraction by ultrasonic waves. Acustica 1990; 70:223–229
7. Follett DH. A versatile schlieren system for beam and waveform visualisation with quantitative real-time profiling capability. In: Evans JA (ed) Physics in medical ultrasound, IPSM Report No. 47, pp 85–93, Institute of Physical Sciences in Medicine, York, UK, 1986
8. Chan HW, Chiang KS, Price DC, Gardner JL, Brinch J. Use of fibre-optic hydrophone in measuring acoustic parameters of high power hyperthermia transducers. Phys Med Biol 1989; 34:1609–1622
9. Bacon DR. Primary calibration of ultrasonic hydrophones using optical interferometry. IEEE Trans Ultrason Ferroelec Freq Contr 1988; UFFC-35:152–161
10. Filipczynski L, Etienne J. Capacitance hydrophones for pressure determination in lithotripsy. Ultrasound Med Biol 1990; 16:157–165

*Ohmic Instruments Co, 102 Chew Avenue, St Michaels, Maryland 21663, USA.

11. Dunn F, Averbuch J, O'Brien WD. A primary method for the determination of ultrasonic intensity with the elastic sphere radiometer. Acustica 1977; 38:58–61
12. Hasegawa T, Yosioka K. Acoustic radiation force on a solid elastic sphere. J Acoust Soc Am 1969; 46:1139–1143
13. Anson LW, Chivers RC. Frequency dependence of the radiation force function (Y_p) for spherical targets for a wide range of materials. J Acoust Soc Am 1981; 69:1618–1623
14. Chivers RC, Anson LW. Choice of target and accuracy of measurement in suspended sphere radiometry. J Acoust Soc Am 1982; 72:1695–1705
15. Reibold R. Calibration of ultrasonic fields using optical holography. Acustica 1980; 46:149–161
16. Zapf TL, Harvey ME, Larsen NT, Stoltenberg RE. Ultrasonic calorimeter for beam power measurements. NBS, US Government Printing Office, Washington DC, USA, Technical Note 686, 1976
17. Fry WJ, Fry WB. Determination of absolute sound levels and acoustic absorption coefficients by thermocouple probes. J Acoust Soc Am 1954; 26:294–317
18. FDA. Guide for measuring and reporting acoustic output of diagnostic ultrasound medical devices. Document 510(k), US Department of Health and Human Services, Food and Drug Administration, HFZ-132, Rockville, MD 20857, USA, 1985
19. NAMAS. The expression of uncertainty in electrical measurements. Guidance document B3003, National Measurement Accreditation Service, National Physical Laboratory, Teddington, Middlesex, UK, 1986
20. Sarvazyan AP, Pasnovkin PN, Shilnikov GV. An extremely simple and rapid method for registration of ultrasound field patterns. Proceedings Ultrasonics International 85, London, 1985, pp 324–328
21. Carlin B. Ultrasonics. McGraw-Hill, London, 1960
22. Bergmann L. Ultrasonics and their scientific and technical applications. G Bell and Sons, London, 1938

4

Hydrophones

S.P. Robinson

4.1 Introduction

Over recent years there has been an increasing requirement for quantitative measurements of the temporal and spatial characteristics of the acoustic fields generated by medical ultrasonic equipment. This has led to the development of miniature piezoelectric hydrophones of varying design, which respond to the pressure in the ultrasonic field.

The ideal hydrophone was introduced in the previous chapter as having the following properties:

1. Small size
2. High sensitivity
3. Omnidirectional
4. Stable
5. Broadband flat frequency response
6. Non-perturbing structure
7. Linear

In practice, it is not possible to manufacture hydrophones with all the above properties, and a suitable compromise has to be accepted. The active elements of ultrasonic hydrophones are typically 0.5 mm or 1.0 mm in diameter, which is comparable to but not smaller than the ultrasonic wavelength. This makes all ultrasonic hydrophones strongly directional and makes alignment crucial when making measurements. Devices with smaller active elements have been developed but these are more difficult to manufacture and generally, the smaller the active element, the lower the sensitivity. Sensitivities are typically 0.1 μV/Pa and preamplifiers are often used both to boost the signal level and to act as an impedance buffer since the hydrophone output impedance is relatively high. Some hydrophone designs are also prone to acoustical resonances, apart from the normal thickness resonance, which make a truly smooth broadband response difficult to achieve. For devices without preamplifiers situated close to the active element, electrical resonances in cables can have a similar adverse effect on the frequency response.

Fig. 4.1 shows examples of the various types of hydrophone which are available. The next section describes the different hydrophone types and gives an assessment of each. The last section describes some of the problems encountered when using a hydrophone to measure the acoustic output of different types of medical ultrasonic equipment. A more detailed description of the performance characteristics needed for a standard measuring hydrophone, is given in the literature [1,2] including details of suppliers [3].

4.2 Types of Hydrophone

4.2.1 Ceramic Hydrophones

The traditional materials used for the active elements of hydrophones have been piezoelectric ceramics, and various commercial devices of this type are

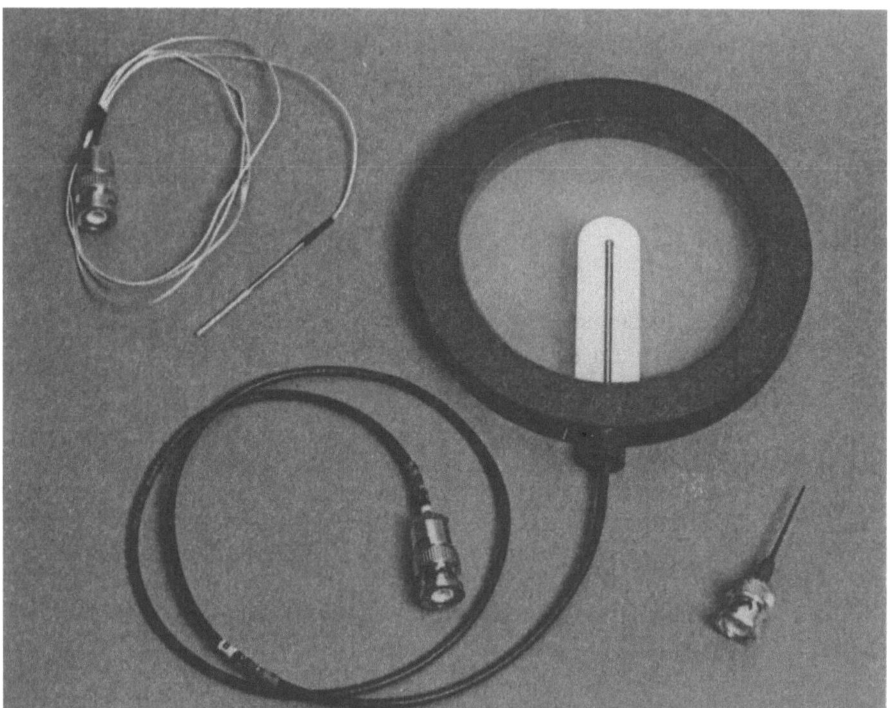

Figure 4.1. Different types of hydrophone. *From left to right*: pvdf needle-probe hydrophone, coplanar shielded membrane hydrophone, ceramic hydrophone.

available. The construction of a ceramic hydrophone is usually in the form of a disc-shaped active element supported at the end of a pencil-like structure made of suitable absorbing material. Fig. 4.2a shows a schematic diagram of a ceramic probe. Radial resonance modes in the active element and reflections in the backing material often lead to unpredictable structure in the frequency and directional responses and therefore to reverberations in the output signal of such devices. There is also evidence to doubt the stability of these devices, and for accurate work they should be calibrated frequently. Until confidence in the stability of a device is gained, it is important to check its calibration each time it is used. These shortcomings limit the accuracy and usefulness of ceramic hydrophones for ultrasonic field characterisation. Nevertheless, they tend to be fairly robust devices and suitable for narrow-band measurements on physiotherapy transducers where often a flat frequency response is not critical.

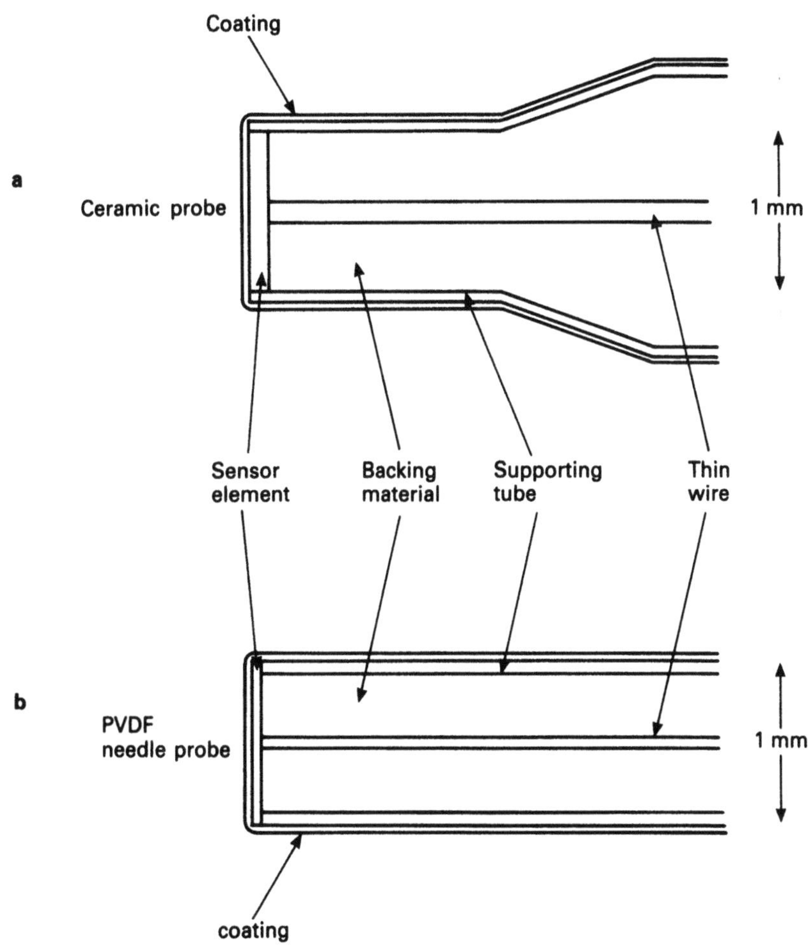

Figure 4.2. Schematic diagram showing **a** a ceramic probe and **b** a pvdf needle-probe hydrophone.

4.2.2 Pvdf Needle-Probe Hydrophones

The piezoelectric polymer polyvinylidene fluoride (pvdf) has been used as an alternative material for the active element of hydrophones (see 2.3.1). In the needle-probe design, shown schematically in Fig. 4.2b, the active element is supported on the end of a needle made from a thin-walled metal tube about 1 mm in diameter. The advantage of pvdf probes over ceramic designs is that the acoustic impedance of pvdf is more closely matched to that of water (giving less pronounced resonances), and experience has shown the devices to be more stable. However, changes in sensitivity of up to 30% have been reported, poor electrical connections frequently being the cause.

Although careful design and construction of pvdf needle-probes has partially overcome the problems of resonance and reverberation which beset ceramic hydrophones, there is still a tendency for rapid fluctuations of sensitivity to occur at frequencies of between 1 and 4 MHz. There is also a rapid fall-off in sensitivity below 1 MHz due to diffraction of the ultrasound around the probe tip.

4.2.3 Pvdf Membrane Hydrophones

A membrane hydrophone consists of a thin sheet of unpoled pvdf film stretched over an annular ring (100 mm internal diameter) with gold/chromium electrodes vacuum deposited on the surfaces. These electrodes overlap only in a small central area, which is then poled and so forms the active element of the device. The diameter of the active element is usually either 1.0 or 0.5 mm. Fig. 4.3 shows a schematic representation of a coplanar shielded type membrane hydrophone, which consists of a single layer of pvdf. There is also a bilaminar shielded version which utilises two layers of pvdf with the active element and almost all of the live electrode sandwiched between two earthed planes. Comprehensive details of the construction and performance of membrane hydrophones is given by Preston et al. [4]. More recently, bilaminar devices have been produced with active elements of diameter 0.1 and 0.2 mm [5].

Membrane hydrophones have the advantage of causing minimal disturbance to the acoustic field since the ultrasonic beam passes through the aperture of the ring in most cases. However, a typical 25 μm pvdf film has an amplitude transmittance of 0.993, 0.92 and 0.78 at 1, 5 and 10 MHz respectively which means that the reflected amplitude (0.089, 0.4 and 0.64 respectively) will become significant in certain applications (see 4.3.4 and 4.3.10).

The frequency response of membrane hydrophones is both broadband and smooth; the fundamental radial resonance mode is that of the internal diameter of the supporting ring which is well below ultrasonic frequencies. The pvdf film will possess a thickness-mode resonance which will occur at a frequency, f, such that:

$$f = c/2t \tag{4.1}$$

where t is the thickness and c is the speed of sound in pvdf. For devices made from 25 μm thick film, this resonance occurs at around 40 MHz for a coplanar and 20 MHz for a bilaminar type as it is double the thickness. Membrane

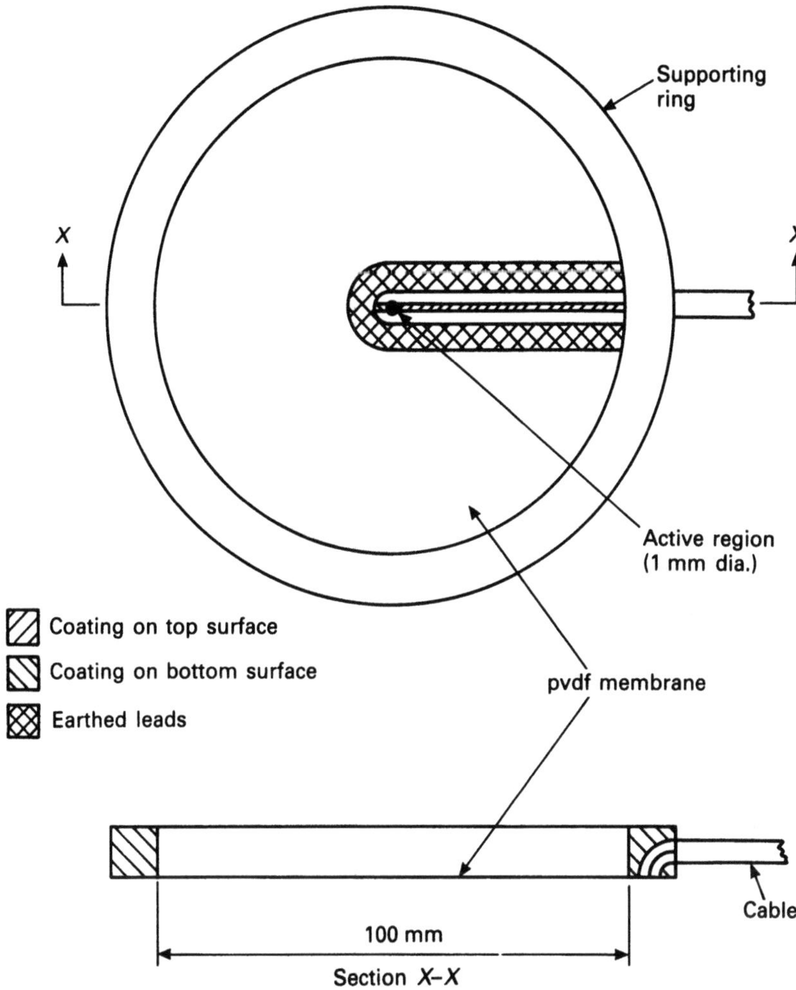

Figure 4.3. Schematic diagram of a pvdf membrane hydrophone.

hydrophones have also been developed with 9 μm thick film and these devices have a correspondingly higher resonant frequency and therefore a higher bandwidth [6]. For example, a 9 μm coplanar shielded hydrophone has a resonance in excess of 90 MHz.

Unlike probe devices, membrane hydrophones do not exhibit rapid fluctuations of sensitivity at lower frequencies. Fig. 4.4 shows a comparison of the frequency response of a pvdf needle-probe hydrophone with that of a 9 μm coplanar shielded membrane hydrophone. The calibrations were performed using the NPL laser interferometer system at small frequency intervals from 0.3 to 15 MHz [7]. As can be seen, the probe exhibits fluctuations of up to 2 dB between 1 and 3 MHz whereas the membrane hydrophone has a "flat" response. This smooth, "flat" response is important when measuring ultrasonic

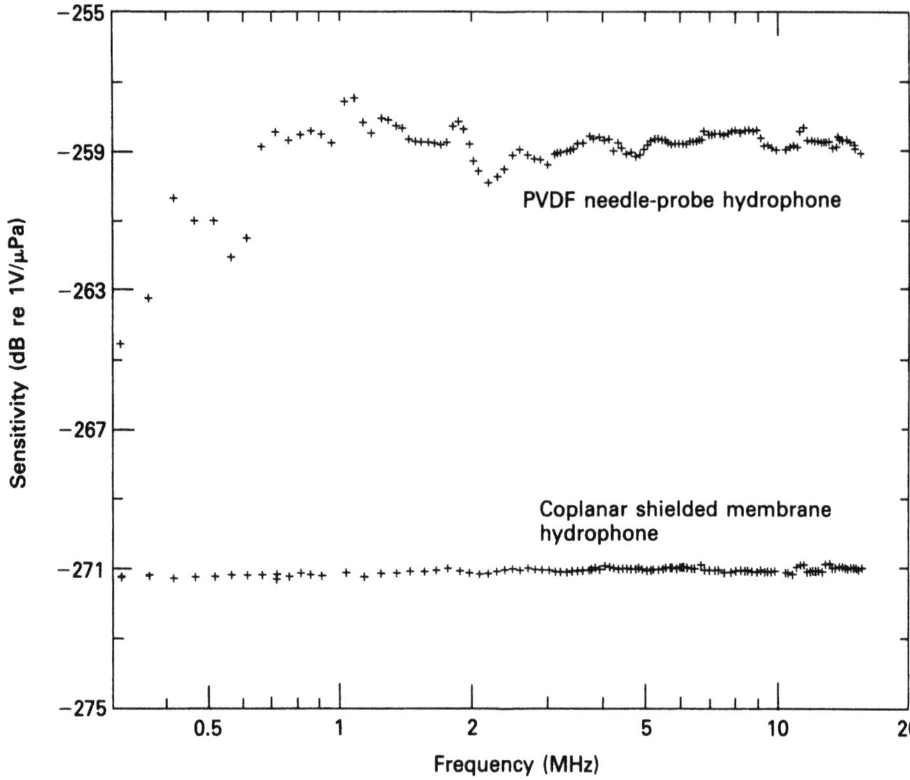

Figure 4.4. A comparison of the frequency response of a needle-probe and a membrane pvdf hydrophone.

fields which exhibit broadband or nonlinear distorted pulse-shapes since the fields contain many harmonic frequency components (see 7.2.1).

As with all ultrasonic hydrophones, membrane hydrophones exhibit strong directionality, and this makes alignment a crucial factor in measurements, especially at high frequencies. Fig. 4.5 shows the typical directional response of a membrane hydrophone at four frequencies. Note that at low frequencies (1 MHz) the hydrophones have large side lobes in their response due to Lamb wave propagation in the film. Probe devices do not suffer from this problem and therefore have superior directional responses at lower frequencies. The angular width of the directional response of a hydrophone is of importance when measuring the field from wide aperture sources, for instance some types of phased arrays and lithotripters.

The properties of the pvdf film have the effect of increasing the effective radius of membrane hydrophones at lower frequencies, and care must be taken when interpreting measurements at these frequencies. This effect is demonstrated by Fig. 4.6 which shows the effective radii for membrane hydrophones with geometrical diameters of 0.5 and 1.0 mm.

Pvdf has been shown to exhibit a linear response to acoustic pressure up to in excess of 60 MPa and membrane hydrophones have been shown to be linear

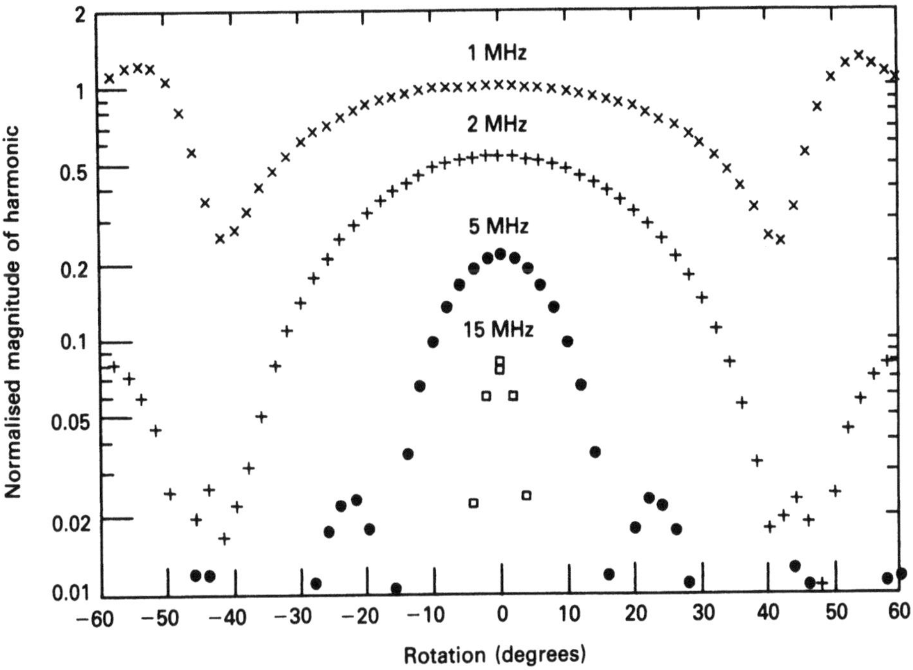

Figure 4.5. Directional response of a typical membrane hydrophone.

up to 10 MPa (± 10%). Membrane hydrophones have stable sensitivities to within a few per cent over 5 years. They are robust devices and have even been used to measure the output of lithotripters which generate acoustic pressures of up to 100 MPa. However, there is evidence that damage to bilaminar devices can occur at the highest acoustic pressures encountered using lithotripters.

4.2.4 Multielement Hydrophones – NPL Ultrasound Beam Calibrator

One advantage of using the concept of an electrode structure deposited on a pvdf membrane is the ability to build a hydrophone with more than one sensing element. Such a device has been developed and is the sensor used in the NPL Ultrasound Beam Calibrator [8].

At the heart of this device is a bilaminar shielded membrane hydrophone which has 21 active elements, each of 0.4 mm diameter and spaced with 0.6 mm between centres. Each element has its own amplification channel which is connected to a fast digitiser, sampling at 60 MHz. Control of the system is achieved using a PC-AT microcomputer. Fig. 4.7 is a schematic diagram which illustrates the concept of the system. With the transducer centred on the hydrophone, the system can select the output signal corresponding to the pressure waveform at any one of the 21 elements and display it in real-time

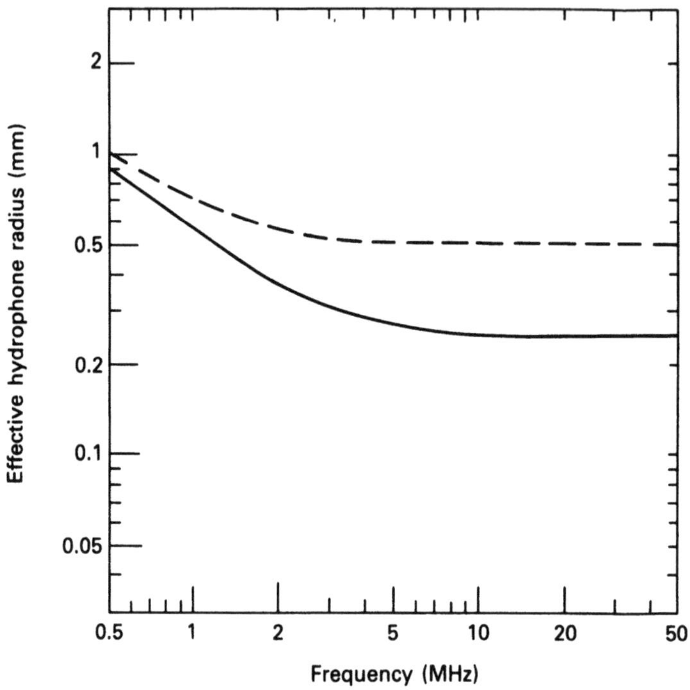

Figure 4.6. The effective radii for membrane hydrophones with geometrical radii of 0.25 mm (*continuous line*) and 0.5 mm (*broken line*) as a function of frequency.

on the graphic display of the PC. Also, the peak voltage measured at each element can be displayed, giving a pressure profile across the transducer beam width. Versatile software can then rapidly calculate the various parameters which need to be measured such as all the spatial and temporal parameters referred to in Chapter 6. This makes the Beam Calibrator a powerful tool when measuring the output of complex systems such as diagnostic scanners.

4.3 Using a Hydrophone

4.3.1 Water

Generally, measurements of the acoustic output of ultrasonic transducers are performed in water, which is a cheap and convenient simulation of human tissue. However, it is preferable that the water is deionised or distilled before use to ensure purity and to reduce the conductivity. Low water conductivity is particularly important for coplanar shielded membrane hydrophones where the active element is exposed to the water. For bilaminar shielded hydrophones, although water conductivity is not a problem and in principle tap water could

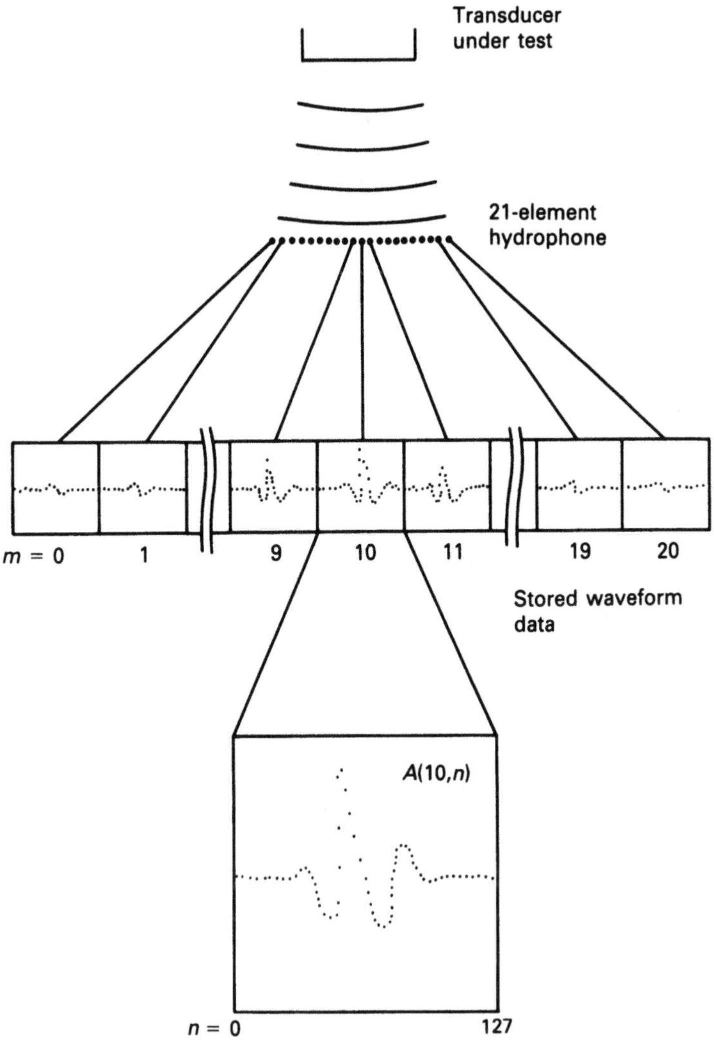

Figure 4.7. Schematic diagram of the NPL Ultrasound Beam Calibrator.

be used, it is still preferable to use pure water in order to prevent the build-up of salt deposits on the surface of the hydrophone.

Severe problems can be caused by air bubbles adhering to the surface of the hydrophone when it is submerged. This problem can be largely eliminated by using degassed water. There are a number of methods used to obtain degassed water [9]. These usually involve either boiling the water for a short time to force dissolved air out of solution, or alternatively, the water is placed under a partial vacuum by means of a vacuum pump. However, degassed water is not always available and so it is advisable to allow the hydrophones to soak for a short time in the water before starting measurements. Bubbles are sometimes trapped under the rim of membrane hydrophones and should be removed with a soft brush. A silvery sheen can sometimes be seen on the hydrophone surface and this is caused by a thin layer of trapped air adhering to the surface. Again, a soft brush, a gentle jet of water or prolonged soaking will remove this.

When making measurements of high power continuous or quasi-continuous wave fields such as those produced by physiotherapy devices, problems can be caused by cavitation. This is when small gas bubbles in the water oscillate in the ultrasonic field, especially if the ultrasonic frequency is near to the resonant frequency of the bubble. This can lead to scattering of the ultrasound which then gives rise to incorrect measurements and can best be eliminated by using degassed water (see Chapter 5).

4.3.2 Mounting and Alignment

Since both the ultrasonic beam from the transducer, and the receiving response of the hydrophone are strongly directional, alignment is crucial in the measurement procedure. In order to align the hydrophone properly, a suitable mount must be used. Membrane hydrophones should be clamped to the mount by their supporting ring with the front of the hydrophone (the side where the ring is almost flush with the membrane) nearest the transducer. Probes should be clamped several centimetres back from the sensitive element.

The hydrophone mount should have five degrees of freedom. It should be possible to move the hydrophone by translation in three orthogonal directions, one of which should be parallel to the direction of propagation of the ultrasound. It should also be possible to rotate and tilt the hydrophone, ideally with the axes of rotation and tilt passing through the active element. Further details about mounting and aligning a hydrophone for maximum signal, are described in Chapter 7.

4.3.3 Shielding

To shield the hydrophone electrically, the earth pin of the hydrophone should be connected to a suitable earth, preferably any metal object in the water tank such as a metal hydrophone mount. Shielded coaxial cable should be used for all signal connections, and using a single common earth connection will help to eliminate pick up at the electrical supply frequency.

4.3.4 Reflections

One source of problems when making measurements is acoustic reflections. These can be reflections from the water surface, from the sides of the water tank, from other objects immersed in the tank, or double reflections from the hydrophone back to the transducer and back again to the hydrophone. These reflections interfere with the direct-path beam and can produce misleading effects and even lead to erroneous measurements. In pulsed or tone burst fields the reflections can be easily observed and therefore isolated since they arrive at the hydrophone after the main direct-path signal.

A useful way of investigating whether a signal is being received directly or via a reflection is to observe the time delay between transducer excitation and received signal and then to change the axial separation of the transducer and hydrophone by moving the hydrophone and note the relative movements of the different signals. Signals received by the hydrophone directly from the transducer will be observed to move at a certain rate away from the transducer excitation signal as the separation is increased. Reflected signals from behind the hydrophone will be seen to move in the opposite direction and double reflections will be seen to move at a higher rate (two or more times faster). If these reflections cannot be isolated it will be necessary to place acoustic absorbers to intercept these stray beams.

Some devices such as physiotherapy transducers, produce continuous wave or very long toneburst (quasi-continuous wave) fields. In these situations, it is not possible to isolate reflections, and acoustic absorbers must be used around the sides of the tank. Good absorbers of ultrasound include butyl rubber and specially-made rubbers which have a characteristic acoustic impedance close to that of water and a high attenuation, often produced by including scattering particles in the rubber. The type of floor mats used in cars (with the pimply surface) and even pieces of carpet have been successfully used. Baffles made from absorbing material are useful, particularly for the water surface.

When using membrane hydrophones in continuous wave fields there can be severe problems from reflections between the transducer face and the hydrophone and these obviously cannot be eliminated by using absorbers. In most cases, the membrane hydrophone can be tilted to destroy the reflections and then measurements can be corrected for the loss of signal if the directional response of the hydrophone and the angle of tilt are known. This process tends to be satisfactory only if the transducer/hydrophone separation is greater than two transducer diameters.

4.3.5 Active Element Size

Errors in measurements of focused fields can also occur due to the hydrophone active element being too large. The hydrophone responds to the pressure averaged over the surface of the element. In fields where the pressure varies rapidly in space, such as highly focused fields, this spatial averaging effect can lead to an underestimate of the pressure. In such fields, the hydrophone with the smallest active element available should be used to reduce this effect (see 7.2.2).

4.3.6 Temperature

The sensitivity of hydrophones can change with temperature. For example, for a coplanar shielded membrane hydrophone, the variation is 0.6% per degree Celsius. Although this variation is small compared to other uncertainties in measurements, efforts should be made to keep the water at a constant temperature. It is important to know the temperature during measurements so that a correction can be applied to the sensitivity figure specified during calibration, often at 20°C.

4.3.7 Bandwidth Limitations

A hydrophone–amplifier combination may have a limited bandwidth which can lead to errors in the measurement of certain parameters, such as peak-positive pressure, when measurements are made on the nonlinear pulsed fields produced by diagnostic equipment (see 7.2.1).

In general, the bandwidth is that of the combined hydrophone and preamplifier. Fig. 4.8 shows the measured frequency response of a 50 μm bilaminar hydrophone (resonance at approx. 23 MHz) and a hydrophone amplifier. Also shown is the resulting frequency response of the hydrophone–amplifier combination. Another factor which can contribute a bandwidth limitation is the frequency response of the oscilloscope or digitiser

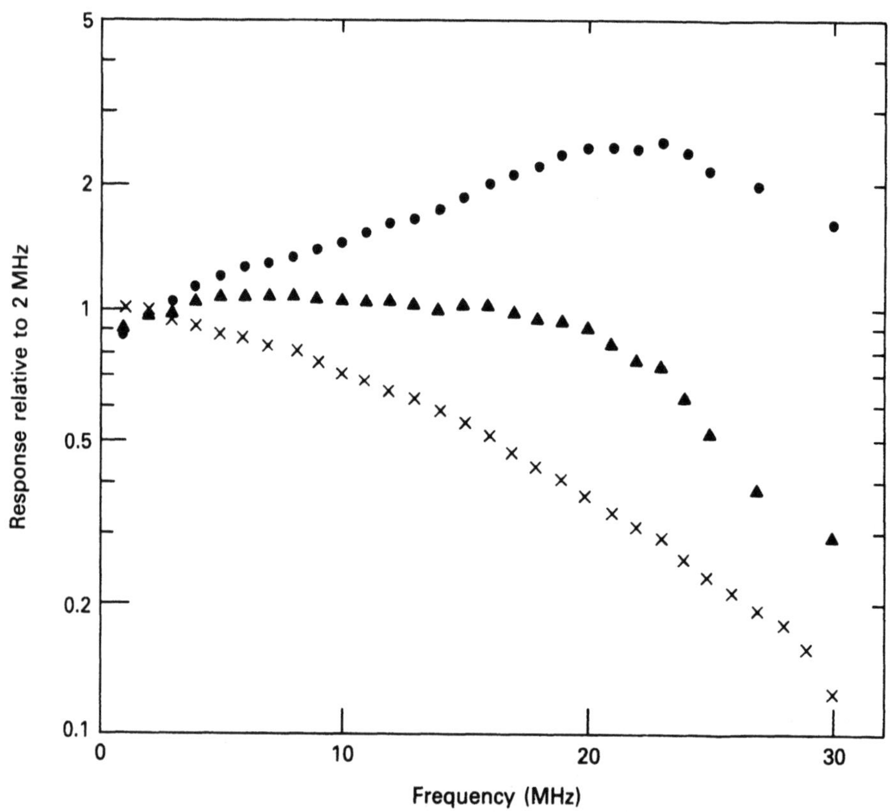

Figure 4.8. The measured frequency response of a 50 μm bilaminar shielded membrane hydrophone (*upper*), a hydrophone amplifier (*lower*), and the hydrophone–amplifier combination (*middle*).

and this must also be taken into account in assessing overall frequency response of a measurement system.

It should also be noted that the hydrophone cable has an associated resonance due to the reflections at the cable terminations, as these are rarely electrically matched. For example, a 60 cm cable has a resonance at 75 MHz and a 15 cm cable is resonant at 300 MHz. Therefore, to achieve a wide bandwidth system, a hydrophone with a high frequency thickness resonance, such as a 9 μm coplanar shielded hydrophone (resonance > 90 MHz), should be used with a short 15 cm cable and a wide bandwidth amplifier (e.g. 75 MHz).

4.3.8 Calibration

Before being used to make measurements, a hydrophone must be calibrated. Without a calibration, it is not possible to determine many of the acoustic parameters which need to be measured when characterising the output of medical ultrasonic equipment. The sensitivity of a hydrophone is the output voltage per unit of applied acoustic pressure and is usually given in either microvolts or nanovolts per pascal, or dB relative to 1 V/μPa.

The calibration of hydrophones may be approached in different ways. Accurate calibration is a specialised, time consuming activity and may be beyond the capability and resource of a test laboratory, medical physics department of a hospital or manufacturer of medical ultrasonic equipment. Fortunately, most suppliers of hydrophones provide calibrations with their devices. For example, all the membrane hydrophones provided by GEC Marconi are calibrated by comparison to a reference hydrophone [10] which has previously been calibrated using the NPL primary standard laser interferometer [11]. The calibration certificate therefore gives traceability to UK National Measurement Standards, as referred to in Chapter 3. Other hydrophones, such as pvdf needle-probe hydrophones from the Medicoteknisk Institut in Denmark, also come with a manufacturer's calibration chart.

Although a hydrophone may have been calibrated at the time of purchase, its sensitivity may change with time and usage. Therefore, it is appropriate here to review the calibration techniques available, and to consider what in-house methods should be implemented to ensure that the calibration is still valid. The two most widely used methods of absolute calibration of hydrophones are reciprocity and planar scanning, both of which are covered by IEC standards [1,12]. Both methods are capable of yielding accuracies of between 5% and 15% over the frequency range 1 to 15 MHz, but to achieve greater accuracy, considerable effort must be devoted to the assessment of systematic errors in the two methods. To avoid these problems, optical interferometry has been developed at NPL which is now capable of achieving accuracies of between 2.5% and 5% over the range of 1 to 15 MHz [11]. However, this method is not simple to implement and is only really suitable for a standards laboratory.

Reciprocity is a relatively simple but powerful technique of which there are several variations. The IEC implementation is called self-reciprocity and involves the use of a transmitting transducer, a reflector and the same transducer used as a receiver. The transmitting current and the received voltage are measured, allowing the acoustic pressure at the position of the reflector to be determined. One assumption made is that the transducer is a reciprocal device, which means that the transmitting and receiving responses are related by a constant factor known as the reciprocity parameter [1]. The reflector is then replaced by the hydrophone to be calibrated and the hydrophone voltage is measured, thereby determining the sensitivity. However, several correction factors must be applied to the sensitivity to account for effects such as diffraction (beam spreading) and electrical loading.

Planar scanning also requires a two-step calibration process [12]. It involves using a hydrophone to scan the field of a transducer, and then performing a spatial integration of the square of the hydrophone voltage. This represents an integration of the square of the acoustic pressure which in turn is proportional to intensity (see Chapter 6). The spatial integral of intensity is equivalent to acoustic power and so if the power transmitted by the transducer is already known, it is possible to "work backwards" and calculate the hydrophone sensitivity. The power radiated by the transducer is usually determined using a radiation force balance (Chapter 5) and the integration of the field is undertaken in a systematic manner based on either raster scans or diametrical scans (Chapter 7). Careful account must be taken of any loss in the integration process due to the scans not covering all of the ultrasonic beam.

Both reciprocity and planar scanning are time consuming and are therefore unsuitable for frequent use. However there remains the need to ensure that the sensitivity of a working hydrophone has not changed and this can be done by establishing a quality regime. One simple procedure that can be adopted is to maintain an in-house reference hydrophone which is not used for routine measurements. The sensitivity of the working hydrophone is then regularly checked against the reference hydrophone by a substitution process in which the signals of the two hydrophones are compared when their active elements are placed sequentially in exactly the same position in the field of a stable source, preferably a plane-piston, unfocused transducer. In addition, to ensure that the reference itself is stable, it is re-calibrated by an absolute method at less frequent intervals, for example, annually. This might be achieved by performing an in-house reciprocity or planar scanning calibration. Alternatively, the reference can be returned to the supplier or submitted to an accredited test laboratory for calibration (see 3.9).

For more information on calibration techniques, the reader is referred to the literature [1,2,9,12,13].

4.3.9 Hydrophone and Amplifier Impedance

The calibration chart which is supplied with the hydrophone will usually state the end-of-cable open-circuit sensitivity of the hydrophone. When in use, there will be some electrical loading of the hydrophone by the amplifier or other measuring device such as an oscilloscope, digitiser or spectrum analyser to which it is connected. The output impedance of a hydrophone is mainly capacitative and for a coplanar shielded membrane hydrophone a large part is due to the water since the active element is not shielded. An important point to note is that the cable also contributes to the capacitance of the device and consequently the device sensitivity will change if the cable length is altered. In this case, it may be necessary to have the device re-calibrated.

To cope with electrical loading effects, corrections based on the ratio of the amplifier and hydrophone capacitances can be applied. If M_c is the end-of-cable open-circuit sensitivity, the end-of-cable sensitivity, M_l, into the electrical load is given by:

$$M_l = M_c C/(C + C_{el}) \tag{4.2}$$

where C_{el} is the input capacitance of the load and C is the end-of-cable capacitance of the hydrophone, including any integral cable and connector. The corrections are usually of the order of 5% for a typical amplifier input capacitance of 5 pF and hydrophone capacitance of approximately 100 pF. However, if the hydrophone has a short cable or is connected directly to an oscilloscope or digitiser, the correction can be much greater.

As has been stated, for accurate measurements the gain and frequency response of the amplifier must be known. Two types of hydrophone amplifier have been designed for membrane hydrophones and are currently commercially available from GEC-Marconi. These are the 5564 and the 560 with bandwidths of 75 and 23 MHz respectively.

4.3.10 Which Hydrophone To Choose?

When choosing a hydrophone for a particular measurement, several factors must be taken into account. The cost of the device and the size of the available budget are perhaps obvious considerations. Other factors depend upon the hydrophone performance properties and the particular type of measurement to be made. Table 4.1 gives information on the three types of pvdf hydrophones which have already been mentioned.

With the information in this table and that contained in the rest of this chapter, it should be possible to make an informed choice as to which type of hydrophone to use. Many practical considerations will enter into the decision. For example, if a supply of deionised water is not readily available, then a coplanar shielded membrane hydrophone would not be a suitable choice. Alternatively, the mounting arrangements might be the limiting factor. It may be that it is easier to build hardware for mounting one type of hydrophone as opposed to another.

Table 4.1. A selection of recommended hydrophones and their properties

Hydrophone type	Element size (mm)	Cable length (mm)	Bandwidth (MHz)	Water required
Bilaminar shielded membrane	0.5 or 1.0	150 750	52 (9 μm) 19 (25 μm)	Any
Coplanar shielded membrane	0.5 or 1.0	150 750	90 (9 μm) 38 (25 μm)	Deionised
Pvdf needle-probe	0.6 or 1.0	1000	15–20	Any

Although the requirements of each measurement must be assessed individually, it is possible to make some broad recommendations.

Pulse Echo – this includes diagnostic imaging and pulsed Doppler.
A 0.5 mm diameter 9 μm thick coplanar shielded membrane hydrophone with a 150 mm cable is the preferred choice. A suitable wideband amplifier should be used, and remember that deionised water is required for a coplanar shielded type. However, for a simpler and a more convenient measurement system, a 0.5 mm diameter bilaminar hydrophone made from 25 μm thick film combined with a 23 MHz bandwidth amplifier (type 560) is recommended.

Continuous Wave – including physiotherapy and cw Doppler.
A 0.6 or 1.0 mm diameter pvdf needle-probe hydrophone is recommended. Alternatively, it is possible to use a membrane hydrophone which has been tilted to destroy reflections (this will require the directional response of the hydrophone to be known). Remember that for physiotherapy measurements, degassed water must be used to prevent cavitation affecting the measurements and damaging the hydrophone.

Lithotripters.
In the case of lithotripters, damage caused by the very high acoustic pressure fields is a major problem for all types of hydrophone. Bilaminar membrane devices with 0.5 mm active elements have been found to be suitable for measurements when the peak pressure does not exceed 100 MPa. However, it is possible for membrane hydrophones to be severely damaged by higher acoustic pressures although the mechanism by which damage occurs is not clear. Needle probe hydrophones also show similar problems of reliability, sometimes surviving well but failing at other times.

There appear to be a number of solutions to the problem of reliability. First, an acoustical attenuator could be used to reduce the acoustic pressure levels at the measuring hydrophone, but this would make predictions to determine the levels without the attenuator difficult because of the problems of nonlinear propagation. Second, the hydrophone could be sandwiched between two thin membranes and the space filled with a fluid such as castor oil which would not readily cavitate. Third, attempts could be made to produce hydrophones at lower cost thereby making failure less important. There will inevitably be developments on these and other fronts. Alternatively, non-invasive optical techniques (see Chapter 3) could be used but as yet such methods are too complex for wide application.

4.4 References

1. IEC. The characteristics and calibration of hydrophones for operation in the frequency range 0.5 MHz to 15 MHz. Document 866 (BS 7041:1989) International Electrotechnical Commission, Geneva, Switzerland, 1987
2. IEC. Measurement and characterisation of ultrasonic fields using hydrophones in the frequency range 0.5 MHz to 15 MHz. To be published as an IEC Standard, Document 87(CO)6, International Electrotechnical Commission, Geneva, Switzerland, 1988

3. IPSM. Guidelines for the routine performance checking of medical ultrasound equipment. Report No. 58, Richardson RE (ed), Institute of Physical Sciences in Medicine, York, UK, 1988
4. Preston, RC, Bacon DR, Livett AJ, Rajendran KJ. Pvdf membrane hydrophone performance properties and their relevance to the measurement of the acoustic output of medical ultrasonic equipment. J Phys E Sci Instrum 1983; 16:786–796
5. Smith RA. The evaluation of small area membrane hydrophones. Report RSA (EXT)2, National Physical Laboratory, Teddington, UK, 1989
6. Bacon DR. Characteristics of a pvdf membrane hydrophone for use in the range 1–100 MHz. IEEE Trans Son Ultrason 1982; SU-29:18–25
7. Robinson SP. A comparison of the frequency response of membrane and needle-probe pvdf hydrophones. In: Evans DH, Martin K (eds). Physics in medical ultrasound II. Report No. 57, pp 79–86, Institute of Physical Sciences in Medicine, York, UK, 1988
8. Preston RC. The NPL ultrasound beam calibrator. IEEE Trans Ultrason Ferroelec Freq Contr 1988; UFFC-35:122–139
9. IEEE 790, 1989. IEEE guide for medical ultrasonic field parameter measurements. Institute of Electrical and Electronic Engineers Inc. New York, USA, 1990
10. Smith RA, Bacon DR. A multiple-frequency hydrophone calibration technique. J Acoust Soc Am 1990; 87:2231–2243
11. Bacon DR. Primary calibration of ultrasonic hydrophones using optical interferometry. IEEE Trans Ultrason Ferroelec Freq Contr 1988; UFFC-35:153–161
12. IEC. The absolute calibration of hydrophones using the planar scanning technique in the frequency range 0.5 to 15 MHz. To be published as an IEC Standard, Document 87(CO)4, International Electrotechnical Commission, Geneva, Switzerland, 1988
13. Harris GR, Lewin PA, Preston RC. IEEE Trans Ultrason Ferroelec Freq Contr, Special issue on ultrasonic exposimetry, 1988; UFFC-35

5

Ultrasonic Power Balances

F. DAVIDSON

5.1 Introduction

Ultrasonic power balances are used to measure the total power of an ultrasonic beam, a parameter which was introduced in Chapters 1, 2 and 3. It is a convenient parameter to measure because many measurement systems are, more or less, portable. The powers from medical ultrasonic equipment range from microwatts, for diagnostic equipment, to several watts for physiotherapy equipment.

There are various reasons for making power measurements but in general they are related to assessment of performance or safety of equipment. For therapy equipment, where continuous-wave or long toneburst transducer excitation is normally used, it is important to know the total power for a number of reasons. One is to ensure the most effective exposure levels are used during patient treatment and another is to ensure exposure does not take place at levels which can be harmful to tissue. It is also useful to know whether the equipment is performing satisfactorily. For diagnostic equipment, where short time-duration pulses of ultrasound are used, power levels are usually low (except for Doppler equipment) which means it is more important to know the spatial distribution of intensity and pressure. Nevertheless, power remains a useful characterising parameter even at these levels of exposure as it can serve as a means of checking consistency when combined with the wide range of field parameters referred to in Chapter 6. Being a simple parameter to measure, power is a useful means of periodic performance evaluation. Owing to the relatively high output levels of some types of Doppler equipment, power is also relevant for the prediction of possible tissue heating.

5.2 The Radiation Force Balance

The most common and convenient way to measure total power is by the detection of the radiation force. Radiation force arises from the transfer of momentum in the ultrasonic beam to a target which intercepts the beam [1,2]. Sometimes the term radiation pressure is used rather than radiation force. Radiation pressure is the force per unit area at a target. Hence, the total force acting on the target, or the radiation force, is obtained by integrating the radiation pressure over the target. Here, we will use the term radiation force as we are only concerned with the total force incident on a particular target.

Radiation force is therefore detected by placing a target in the beam of an ultrasonic transducer and using some sort of sensor attached to the target to indicate the force. It can be shown that the force is directly proportional to the ultrasonic power providing the whole beam is intercepted [3]. In Chapter 3 the equation for radiation force, F, was given as:

$$F = \frac{hW}{c} \tag{5.1}$$

where W is the total power and c is the speed of sound in the medium. The factor h depends on the type of target used (types of target and their associated h value will be discussed in 5.3). As the speed of sound, c, is temperature dependent (see Chapter 6), it is necessary to monitor the temperature of the measurement medium.

Eq. (5.1) assumes plane waves are incident on the target which is a reasonable assumption for fields generated by plane piston transducers. However, focused transducers are commonly encountered, in fact the majority of medical ultrasonic transducers use some form of focusing. To account for the range of angles of incidence of a focused beam on a target, a modified version of Eq. 5.1 can be used. It has been shown [4] that for a focused transducer of radius of curvature R and active element of radius a_1, the relation between the radiation force and the ultrasonic power is:

$$F = \frac{hW}{c}\,\frac{(1 + \cos\beta)}{2}$$

where β = arc sin a_1/R. If the radius of curvature is not known or the focusing is achieved using phased excitation of a group of elements, such as in a phased array or linear array, the radius of curvature can be assumed to be the distance from the transducer face to the focus.

To decide upon a suitable measurement system, the power range to be measured must be considered. For diagnostic equipment, the total power is usually in the microwatt to milliwatt range, although Doppler equipment may generate levels as high as hundreds of milliwatts. To measure very low power levels, the system will require to be extremely sensitive while also able to cater for transducers with large front-faces. Another factor to be considered when deciding on a measurement system is portability. It is difficult to produce a portable device which has high sensitivity, hence the most sensitive radiation balances tend to be laboratory-based systems (see 5.4). If a portable device is essential it may not be possible to measure reliably powers below 1 mW.

To measure in the power range from milliwatts to several watts, the way that most systems work is to measure the displacement of the target or float (assuming that displacement is proportional to power) or to measure the restoring force necessary to keep the target fixed. From Eq. (5.1), it follows that power, W, is found from:

$$W = cF/h \tag{5.2}$$

If the target-type factor, h, is unity, as it often is, this means that the radiation force per unit power is only 0.69 μN/mW. With such a low sensitivity, balance targets have to be kept as small as possible to avoid too many problems from instabilities. Hence, a balance with a target sufficiently large to detect the total power from a 100 mm wide linear array is normally unstable and therefore has a high threshold for detectability. In general, it is difficult to measure total powers of less than 1 mW with a balance using a target larger than 50 mm diameter. Unfortunately, the total power output of many scanning systems is in the milliwatt range and therefore power measurement in scanning mode is often difficult (see 5.5.5).

5.3 Choosing a Configuration and Target

There are various configurations of target and transducer for radiation balances and the common ones are shown in Fig. 5.1 [5]. Either the target or the transducer can be kept fixed and the force detected at the non-fixed element. However, in practice, the attachment of the transducer cable tends to apply constraining forces or restrict its free movement. Hence, it is usual to have a fixed transducer and a non-fixed target.

In order to determine the total acoustic power output of a transducer it is essential that the target be of sufficient size to intercept the whole of the ultrasonic beam. Although the criterion for total interception of the beam is complex, as it depends on the particular geometry of the ultrasonic field, simple guidelines can be given. For instance, in the case of plane piston transducers it has been shown [3] that the target diameter should be at least double the diameter of the active element of the transducer. As the target can be made of either reflecting or absorbing material, all the ultrasound must therefore be either reflected or absorbed by the target. In fact, an additional requirement is that the angle of incidence of the beam on the target must be constant over the area of the beam. If the beam being measured is divergent or is a sum of individual beams in different directions, such as from phased arrays or sector scanners, different parts of the beam or different acoustic pulses will impinge on the target surface at different angles. As the radiation force balance detects the momentum in one direction, a cosine error will be introduced across the beam or beams dependent on the angle of incidence (see 5.5.5).

There are basically two types of detector. A compensating detector will apply a counter-acting force equivalent to the radiation force, so that the target will effectively not move. In a non-compensating system the target will be allowed to move and its displacement will be related (ideally proportional) to the radiation force.

5.3.1 Reflecting Targets

Reflecting targets of various shapes may be used (see Fig. 5.1). In all cases, the radiation force acting on the target has two components: the first component is the reaction to the momentum of the incident ultrasound beam, and the second is the reaction to the momentum of the reflected beam which in turn depends on the reflectivity of the target. It is the direction and magnitude of the reflected beam which determines the value of the parameter h in Eq. (5.1). Most reflecting targets are made of thin metal and are air-backed, thus ensuring the reflection is 100%. Apart from the net force in the direction of the incident ultrasound beam, a reflecting target will also experience a lateral force. If the

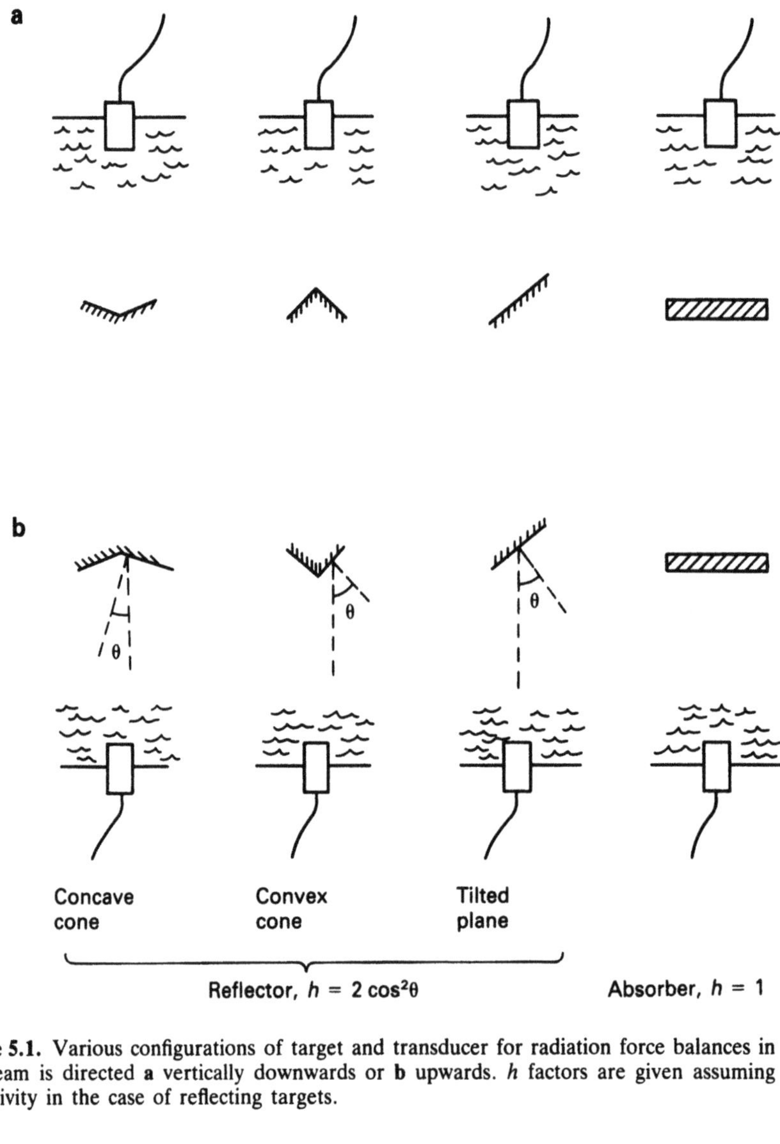

Figure 5.1. Various configurations of target and transducer for radiation force balances in which the beam is directed **a** vertically downwards or **b** upwards. h factors are given assuming 100% reflectivity in the case of reflecting targets.

reflecting target is not constrained in this direction then it may move. One geometry of target which overcomes this problem is the unrestrained concave cone which tends to be self-centring.

Looking at each type of reflecting target shown in Fig. 5.1, an air-backed concave cone has an h factor of $2\cos^2\theta$, where θ is the angle between the direction of the incident ultrasound and the normal to the reflecting surface. As already mentioned, one useful feature of such a target is that it will be self-centring. However, a less desirable feature is that standing waves may be set up between the target and the transducer if their separation is small. This can affect the transducer emission.

For an air-backed convex cone the h factor is still $2\cos^2\theta$. Here, standing waves are not a problem providing the cone angle is sufficiently small. Obviously, a very obtuse cone would reflect ultrasound back to the transducer. One configuration which is widely used is a right-angled cone such that the target surface intercepting the beam is at 45° to the beam axis. This ensures that the reflected ultrasound is directed perpendicular to the beam axis and therefore away from the transducer to the sides of the vessel, which are lined with an acoustic absorber. However, the target is not self-centring. This has to be overcome by the operator carefully aligning the transducer and target before starting measurements. However, if the target is suspended from one arm of a microbalance there will be a gravitational restoring force which will help in maintaining horizontal stability of the target.

The tilted plane reflector is also not self-centring, but it has the advantage of being able to cope with larger transducer beams.

As all the reflecting targets absorb only a small proportion of the acoustic energy, no buoyancy changes due to thermal expansion of the target will take place. Hence thermal drifts caused by absorption have no effect on the force being measured (see 5.5.4). However, as the ultrasound has ultimately to be absorbed somewhere, usually at the absorbing lining of the vessel walls, convection currents can often be set up in the water which can then produce drifts in the measurements.

5.3.2 Absorbing Targets

It might at first appear that an absorbing target makes the measuring system simpler because it does not have some of the disadvantages noted in the previous section for reflecting targets. However, absorbing targets introduce different problems of measurement. Although total absorption corresponds to an h factor of one, it is difficult to produce a material that will completely absorb incident ultrasound with no reflections. More importantly, the absorbed ultrasound will cause the target material to heat up resulting in thermal expansion and a change in buoyancy. This in turn may cause the weight of the target to drift and give rise to significant errors in the measurement of the radiation force. Work is currently in progress to produce alternative absorbers and to minimise buoyancy changes whilst maintaining a simple system.

5.4 Examples of Radiation Force Balances

5.4.1 General

There have been many practical designs of radiation force balance, each one being designed to realise different objectives. Early designs of balance aimed at measuring low powers (diagnostic beams) used either reflecting or absorbing targets and were mainly intended for laboratory use [6–12]. Other designs were aimed at measuring the higher powers generated by therapy transducers [13–15] or have been designed with a large dynamic range in mind [16]. More recently there has been considerable interest in the development of portable radiation force balances [17–21]. A useful summary and assessment of different types of radiation force balance and commercially-available systems is given in the IEEE Standard [22].

Farmery and Whittingham [17] produced a neat system consisting of a flat reflecting target mounted on a moving-coil arrangement with a feed-back magnetic restoring force to maintain the target position fixed. However, the balance compartment was filled with liquid paraffin which introduced a frequency dependence in the sensitivity of the balance to incident ultrasonic power, caused by absorption in the oil. Duck et al. [20] produced a similar system differing mainly in having a reduced path length between the entry port membrane and the reflecting vane, thereby reducing the absorption losses in the oil. More recently, Perkins [21] has developed a versatile force balance using a vertical support arrangement for a large reflecting target. Unlike the two previous designs, the liquid filling the target chamber is water and additionally the entry port membrane can if necessary be removed and the whole balance immersed in water. The design of the balance copes with powers from a few milliwatts to watts and is therefore suitable for both diagnostic and therapy beams.

The following three sections describe in greater detail three designs of radiation force balance which illustrate the three main features described above: a laboratory-based high sensitivity balance for low power measurements, a small portable balance for diagnostic powers and a semi-portable radiometer for therapy power levels.

5.4.2 The NPL Radiation Force Balance

For measurement of total power in the range below 200 mW, the radiation force balance shown in Fig. 5.2 is used at NPL [23]. This is an example of a compensating type of balance. It is based on the use of a sensitive servo-controlled microbalance to detect the force on the target. Because of the high sensitivity of this balance, extra precautions are needed against air-currents and the water tank requires a draught shield.

A convex, conical, air-backed, reflecting target is suspended in the tank from one arm of a microbalance. The microbalance registers the radiation force on the target as an apparent change of target weight. The feedback system of the microbalance causes the servo to produce a voltage which is proportional to the applied force and therefore to the transducer output power.

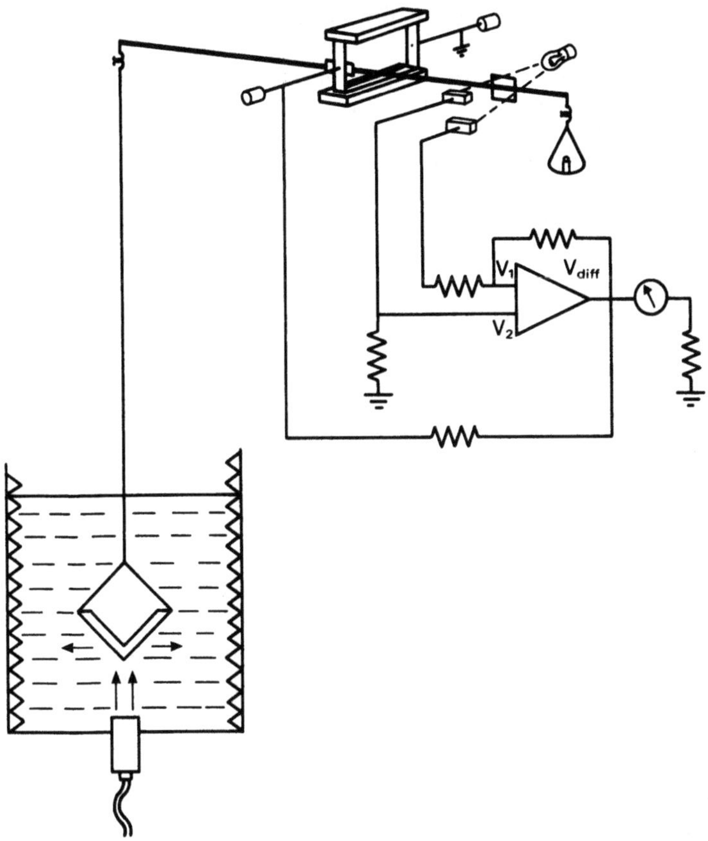

Figure 5.2. Schematic diagram of the NPL radiation force balance.

The tank is filled with degassed water. The walls of the tank are lined with an acoustic absorber, e.g. carpet or rubber, which must be soaked to ensure thorough wetting so that no pockets of air are trapped. The transducer is fixed to the bottom of the tank and the beam is directed vertically upward. It is essential that the float is centred above the transducer accurately by the operator as this is not a self-centring target (see 5.3.1).

In this system, the total beam power is measured for a particular transducer drive voltage. The radiation conductance (power divided by the square of the drive voltage) may then be calculated. Measuring the radiation conductance for different drive voltages is a useful check for consistency of the transducer performance as it should be independent of drive voltage.

Although suitable for accommodating transducers with cylindrical bodies, it is difficult to measure the output power of many commercial transducers as they often have non-cylindrical casings. Hence, the main application of this balance at NPL is that of being a primary standard for the measurement of ultrasonic power.

5.4.3 A Commercial Portable Power Balance

The Doptek power balance of design similar to that published by Cornhill [19] measures in a range similar to the NPL radiation force balance, approximately up to 400 mW. Fig. 5.3 shows the important features of this balance, which uses an absorbing float, held in position magnetically and employs the force-feedback principle. The radiation force here is proportional to the change in coil current. A membrane encloses the cavity in which the float is suspended and it is filled with water. Above the membrane a coupling medium (usually degassed water) is placed in the cavity and the transducer is aligned vertically. The important feature of this system is that it is portable and simple to use. A restriction is that the diameter of the entry port aperture is not large enough to accept most linear arrays and large diameter transducers.

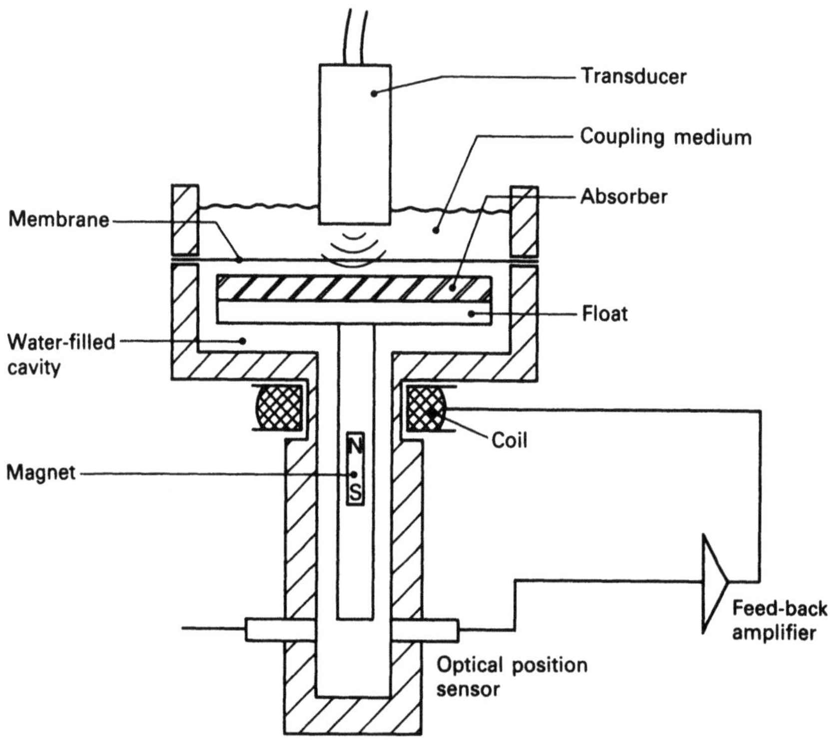

Figure 5.3. Schematic diagram of the Doptek Balance.

5.4.4 The Tethered Float Radiometer

A very simple, non-compensating system is the tethered float radiometer [15], shown in Fig. 5.4, which has a power measurement range of 200 mW to 9 W, suitable for therapy transducers.

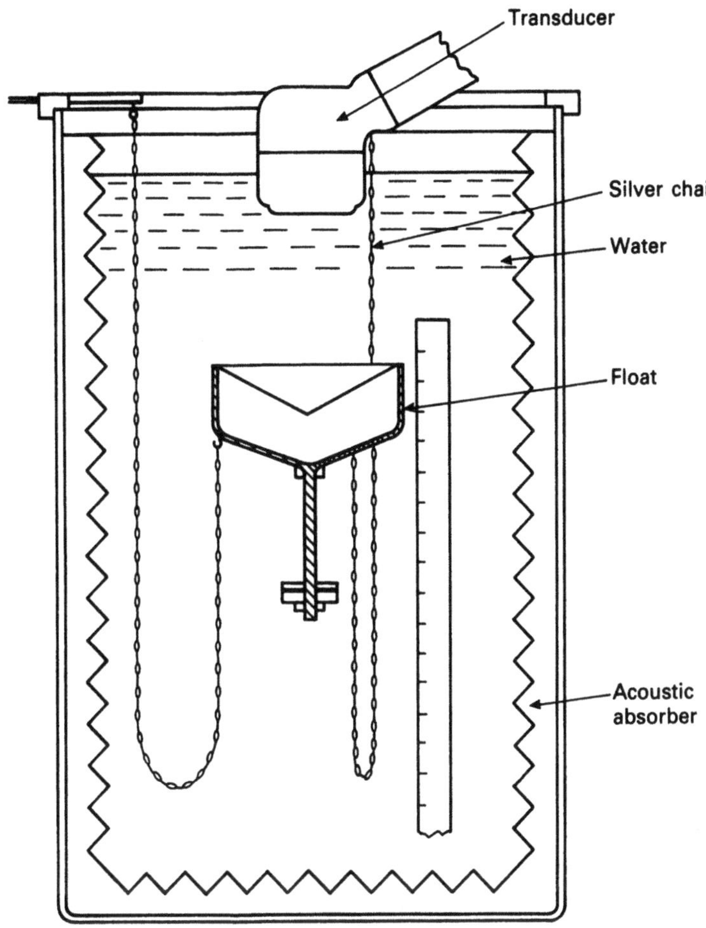

Figure 5.4. Schematic diagram of the tethered float radiometer.

The target here is a buoyant float with concave reflecting surface. It is attached by three silver chains of equal length to the top of the water bath, allowing part of the chain to hang below the float. The weight of chain borne by the float depends on the vertical position of the float and the weight in water per unit length of the chains. By adjusting the buoyancy of the float, it can be made to come to rest at any vertical position.

When the transducer is turned on the radiation force causes the float to be pushed downwards. This movement reduces the length of chain – and therefore the weight of chain – supported by the float as more of the chains become supported by the rim of the water bath. A ruled scale is used to measure the equilibrium deflection. To improve accuracy, a travelling microscope may be used to measure the deflection.

The sensitivity of the radiometer to incident ultrasonic power depends on the weight per unit length of the chains in water and it is possible to choose different types of chain to give different sensitivities. Typically, a sensitivity of 10 to 30 mm/W is possible.

The water bath is filled with degassed water and lined with an acoustic absorbing material such as unlooped carpet or spikey-surfaced rubber such as a car mat. The transducer is mounted by a clamp on a stand and should be placed centrally, directly over the float. The water temperature is measured, as both the speed of sound and the attenuation are temperature dependent. For a specific transducer output, a series of repeat measurements to determine the deflection is undertaken which allows the random uncertainty to be calculated. If measurements are made at fixed time intervals after turning the transducer on or off, it is possible to account for any drift (see 5.5.4).

From the measured displacement the total power is calculated. If the transducer drive-voltage is known, the power per volts squared (radiation conductance) can also be calculated. The total power, W, in watts, is given by:

$$W = \frac{S \, \Delta x \, c}{2 \cos^2 \theta} \tag{5.3}$$

where Δx is the displacement of the target, S is the sensitivity factor (force required to displace the target by unit length), θ is the angle between the normal to the reflecting surface of the cone and the direction of incidence of the beam (cone half angle is equal to $90° - \theta$) and c is the speed of sound in the water. The sensitivity is determined by placing known weights (in water) on the target and noting the displacement. It is also possible to derive the sensitivity knowing the weight per unit length of the chains but this method is less accurate because of the uncertainty in determining the buoyancy correction for the chains. The speed of sound is temperature-dependent and values for c are given in Chapter 6.

5.5 Measurement Considerations

5.5.1 Propagating Medium and Attenuation

Ideally, the medium in which ultrasonic power is measured should be analogous to human tissue, and with the same characteristic acoustic impedance (ρc). In practice, water is a convenient medium whose acoustic impedance is close to that of tissue. Alternatives are sometimes considered consisting of solutions and mixtures of chemicals, but these tend to introduce new factors which complicate the measurements and reduce accuracy. These complications may arise from the limitations of maintaining a set concentration level and ensuring that the ultrasound does not cause any reactions in the medium.

As ultrasound travels through a medium it is attenuated, i.e. some of its energy, and therefore its momentum, is transferred to the medium. This attenuation is frequency dependent and has greatest effect at the higher frequencies. There are two factors of note here:

1. Power loss from the beam increases as the separation between transducer and target increases. This loss is calculable using the known attenuation of

water [24] in combination with the known distance and frequency. To correct for attenuation, the transducer output power, W_c, is determined from:

$$W_c = W \exp(2\alpha d) \tag{5.4}$$

where W is the measured power, d is the depth of the float below the transducer (in metres) and α is the amplitude attenuation coefficient (in metre^{-1}) of water at the appropriate frequency. It is useful to remember that the attenuation of water is inversely proportional to the square of the frequency. To check the consistency of the attenuation correction, measurements can be repeated at the same drive power but different transducer–target separations. After applying the exponential attenuation correction to the measurements at each distance, the calculated power emitted by the transducer should be constant.

2. Attenuation in water is lower than in tissue and an adjustment factor can be applied when analysing the results in order to estimate the power in an ultrasound beam at a particular clinical site. This may be necessary when considering the significance of thermal effects from an ultrasound beam. In this case, the power, W_t, in tissue is calculated from the transducer power, W_c, using:

$$W_t = W_c \exp(-2\alpha_t d_t) \tag{5.5}$$

where α_t and d_t are the amplitude attenuation coefficient and overall thickness respectively of the tissue between the transducer face and the point of interest. Of course, the tissue layer could be a composite of layers of different tissue.

It is important to note that, for both the above correction procedures, linear propagation has been assumed. In practice, extra loss occurs in many beams owing to a phenomenon called "nonlinear loss" [25].

5.5.2 Streaming and Convection Currents

Streaming and convection currents may be set up which have the effect of apparently increasing the measured power. To prevent the streaming currents being set up, an acoustically transparent membrane is inserted between the transducer and the target. As an additional precaution, the membrane can be tilted at an angle other than perpendicular to the beam-axis. This is to avoid standing waves being set up between the membrane and the transducer. However, some systems for measuring total power have membranes which are exactly perpendicular to the transducer beam and it is important to be aware that these could produce standing waves which may significantly affect the measurements.

Convection currents can arise from the heating of the transducer whilst in operation (they are not 100% efficient) and the transfer of heat to the surrounding medium. In addition, convection currents are caused by heating of the medium (water) as a result of the ultrasound being partially absorbed. If the membrane is placed close to the target this will reduce the effects of both the streaming and convection currents.

5.5.3 Cavitation

For power measurements the water must be degassed to avoid the possibility of cavitation occurring. Cavitation, briefly, is the formation or activation of small gas bubbles in the ultrasound beam. These bubbles have various effects on the measurements but mainly they produce an apparent increase in attenuation of the ultrasound. It is essential to look for evidence of cavitation during the measurement process as air will be gradually absorbed into the degassed water which could then lead to the formation of bubbles. Problems caused by cavitation are usually most severe when making measurements on physiotherapy equipment, especially at 1 MHz and below. If the transducer is immersed in water with its flat face pointing downwards, bubbles can often build up on the face without the operator realising. This is avoided by frequently observing the transducer front face. Many situations in which lack of reproducibility of measurements occurs are attributable to the presence of bubbles.

5.5.4 Thermal Drifts

Most radiation force balances suffer from drifts in their "zero" levels, either a readout level or a target position. The drift is usually due to fluctuations in ambient temperature or draughts, or to heating of the target (if the target is an absorber), the water, the tank absorbing lining (if the target is a reflector) or the transducer itself. The rate of drift may therefore be different between the transducer being permanently switched off and the rate observed when the transducer is switched on.

Although it is advisable to undertake special investigations to isolate the source of the drift, in general the following guidelines should be followed:

1. Always try to minimise the time elapsed between turning on the transducer and making a measurement of the change in balance reading (the same applies to turning it off)
2. Take mean values of the change in balance reading for the ON–to–OFF and OFF–to–ON conditions
3. Consider extrapolation of the change in readings caused by thermal drift to a common time in the measurement sequence (see the example shown in Fig. 5.5).

5.5.5 Coping with Scanning Systems

The total output power of many scanning systems is in the milliwatt range and the transducers are often very large (e.g. linear arrays). Nevertheless, it is often possible to measure the power in a static beam mode such as M-mode from which the power in imaging mode can then be derived. To do this it is necessary to know the pulse repetition rates and to assume the pulse waveforms are identical in the two modes. If *prr* is the pulse repetition rate in M-mode and W_m is its total power, then the beam power, the power in one scan line,

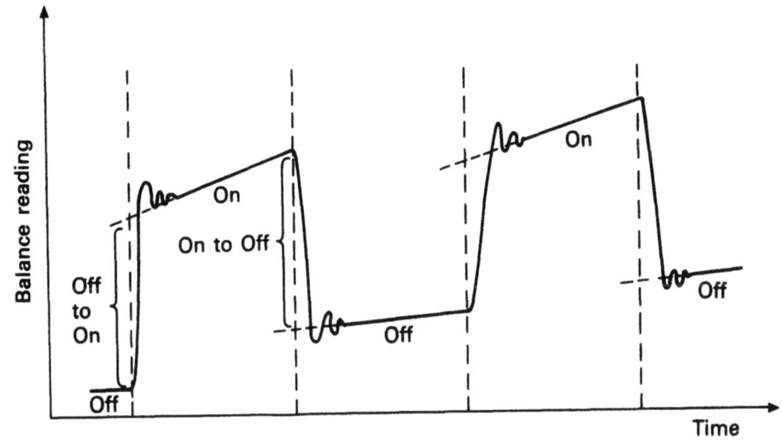

Figure 5.5. Illustration of the effects of thermal (and other) drifts and one way to minimise errors by extrapolating On and Off balance readings to common times.

will be W_m/prr. If there are N scan lines then the total power emitted by the array in imaging mode will be:

$$N \cdot \frac{W_m}{prr} \tag{5.6}$$

Although it is theoretically possible to determine the power in this manner, it may not be possible to couple physically the imaging transducer to the radiation balance. Even if the target of a balance is large enough, the entry port of the balance must accept the transducer and this may be difficult for a linear array which will probably be bigger than the entry port dimensions. If the entry port of the balance is horizontal then it is sometimes possible to couple the transducer utilising a water meniscus as shown in Fig. 5.6.

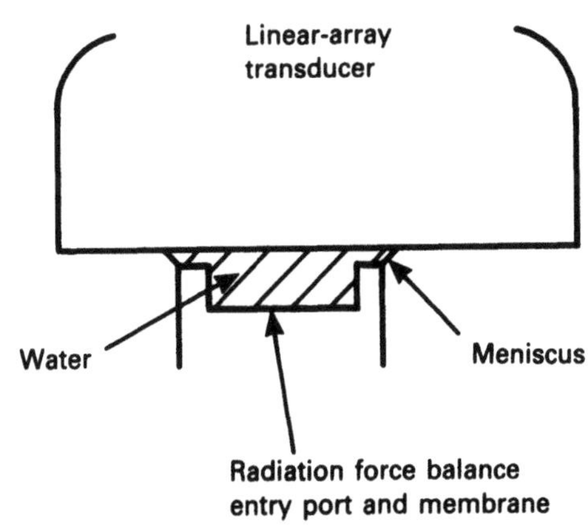

Figure 5.6. Coupling a linear array transducer to a radiation force balance using a horizontal entry port.

Just as the linear array cannot normally be accommodated, neither can a divergent or fan beam such as that produced by a sector phased array or by a convex linear array transducer. As the radiation force balance responds to the force in one direction only, the response to a fan beam will vary across the beam as each portion of the beam intercepts the target at a different angle, introducing a cosine dependence. The angle of incidence of the beam on the target surface will therefore have to be known if an attempt is to be made at applying a correction for this effect. For a reflecting target, the angle of incidence of each element of a fan beam on the target will vary across the beam. Unless the geometry of the beam in relation to the target surface is known it may not be possible to correct for this effect. For an absorbing target, a correction can be made using the following method.

Consider the beam made up of the different radial scan lines and let θ be the angle between any line and the direction of sensitivity for the balance (normally vertical). For a uniform set of N ultrasonic lines in a fan-form, each one of power W_i, the total power measured using an absorbing target will be:

$$W_m = \sum_{i=1}^{i=N} W_i \cos \theta_i \tag{5.7}$$

whereas the total power emitted will be:

$$\sum_{i=1}^{i=N} W_i$$

If each scan line emits the same power, W (where $W = W_i$), the total power, NW, is given from Eq. (5.7) by:

$$N W = \frac{N W_m}{\sum_{i=1}^{i=N} \cos \theta_i} \tag{5.8}$$

Knowing the geometry of the scan it should then be possible to correct a measurement made on a fan-shaped beam using Eq. (5.8).

Applying this concept to a reflecting target is in theory possible but the angle of intersection depends on a number of geometrical factors, especially for convex or concave conical reflectors. Applying corrections in these cases will therefore be very difficult. Often it is easier to determine total power by using a hydrophone and integrating the square of the hydrophone voltage, see Chapters 6 and 9.

5.6 Summary of Important Points to Note During Measurements

The following notes summarise some of the important general points to be aware of during the process of making measurements of ultrasonic power using radiation force balances.

1. It is essential before commencing measurements that all air-bubbles are removed from the target surface, the transducer-face and from the lining of the vessel
2. The medium in which the measurements are to be made (normally water) should be degassed to avoid cavitation occurring
3. Where necessary, the transducer connections should be water-proofed before immersing the transducer
4. Reflected ultrasound must be wholely absorbed elsewhere, i.e. by lining the vessel walls
5. For a concave reflecting target, ensure adequate transducer–target separation to avoid interference effects with the transducer
6. Repetition of measurements not only allows the determination of random uncertainties, but following a time sequence can help to reduce the effects from thermal drift
7. Take account of streaming and convection currents at higher frequencies by introducing a membrane close to the target

A final additional check on the performance of the apparatus is to repeat measurements at different transducer–target separations. The same total power value should be obtained after applying the appropriate attenuation corrections.

5.7 References

1. Livett AJ, Emery EW, Leeman S. Acoustic radiation pressure, J Sound and Vibration 1981; 76:1–11
2. Torr GR. The acoustic radiation force. Am J Phys 1984; 52:402–408
3. Beissner K. Minimum target size in radiation force measurements. J Acoust Soc Am 1984; 76:1505–1510
4. Beissner K. Radiation force calculations. Acustica 1987; 62:255–263
5. Preston RC. Measurement and characterisation of the acoustic output of medical ultrasonic equipment Part 2. Med and Biol Eng and Comput 1986; 24:225–234
6. Carson PL, Fischella PR, Oughton TV. Ultrasonic power and intensities produced by diagnostic ultrasound equipment. Ultrasound Med Biol 1978; 3:341–350
7. Hill CR. Calibration of ultrasonic beams for biomedical applications. Phys Med Biol 1970; 15:241–248
8. Kossoff G. Balance technique for the measurement of very low ultrasonic power outputs. J Acoust Soc Am 1965; 38:880–881
9. Rooney JA. Determination of acoustic power outputs in the microwatt–milliwatt range. Ultrasound Med Biol 1973; 1:13–16
10. Wells PNT, Bullen MA, Follett DH, Freundlich HF, Angell James J. The dosimetry of small ultrasonic beams. Ultrasonics 1963; 1:106–110
11. Wells PNT, Bullen MA, Freundlich HF. Milliwatt ultrasonic radiometry. Ultrasonics 1964; 2:124–128
12. Wemlen A. A milliwatt ultrasonic servocontrolled balance. Med Biol Eng 1968; 6:159–165
13. Kossoff G. Calibration of ultrasonic therapy equipment. Acustica 1962; 12:84–90
14. Lunt MJ, Ashley B. A simple radiation balance for measuring ultrasonic power. J Med Eng Tech 1979; 3:194–197
15. Shotton KC. A tethered float radiometer for measuring the output from therapy equipment. Ultrasound Med Biol 1980; 6:131–133
16. Anson LW, Chivers RC, Adach J. Ultrasonic radiation force devices with non-linear mechanical suspensions. Acustica 1989; 67:226–234

17. Farmery MJ, Whittingham TA. A portable radiation-force balance for use with diagnostic ultrasonic equipment. Ultrasound Med Biol 1978; 3:373–379
18. Bindal VN, Singh VR, Singh G. Acoustic power measurement of medical ultrasonic probes using a strain gauge technique. Ultrasonics 1980; 18:28–32
19. Cornhill CV. Improvement of portable radiation force balance design. Ultrasonics 1982; 20:282–284
20. Duck FA, Starritt HC, Aindow JD, Perkins MA, Hawkins AJ. The output of pulse-echo ultrasound equipment: a survey of powers, pressures and intensities. Brit J Radiol 1985; 58:989–1001
21. Perkins MA. A versatile force balance for ultrasound power measurement. Phys Med Biol 1989; 34:1645–1651
22. IEEE 790:1989. IEEE guide for medical ultrasonic field parameter measurements, Institute of Electrical and Electronic Engineers Inc. New York, USA, 1990
23. Livett AJ, Leeman S. Radiation pressure and its measurement, Proceedings of the 1983 IEEE Ultrasonics Symposium, p 749
24. Pinkerton JMM. A pulse method for the measurement of ultrasonic absorption in liquids: results in water. Nature 1947; 160:128–129
25. Duck FA, Perkins MA. Amplitude dependent losses in ultrasound exposure measurement. IEEE Trans Ultrason Ferroelec Freq Contr 1988; 35:232–241

6

Hydrophone-Based Measurements on a Specific Acoustic Pulse Part 1: Field Characterisation

R.C. PRESTON

6.1 Introduction

This chapter deals with the use of a hydrophone to measure and characterise a single acoustic pulse or a sequence of identical acoustic pulses generated by an ultrasonic transducer. Later chapters deal with the extension of these basic techniques to cope with more complex ultrasonic fields such as those generated by ultrasonic scanners.

It is assumed here that the hydrophone responds to the acoustic field so that the output voltage of the hydrophone as a function of time, $V(t)$, is proportional to the acoustic pressure $p(t)$ at the active element of the hydrophone. Therefore, reference to waveforms will ignore problems such as the frequency response of the hydrophone/measurement system, spatial averaging etc. which are dealt with in Chapter 7. It is also assumed that the most appropriate hydrophone has been chosen for the particular measurements.

6.2 The Acoustic Pulse Waveform

Assuming that the transducer generating the ultrasound is excited with a single electrical pulse, Fig. 6.1 shows a typical hydrophone signal that might be measured in the field.

Whilst such a diagrammatic representation, assuming the vertical and horizontal axes are scaled in absolute units of acoustic pressure and time respectively, provides all the relevant information about the acoustic pulse, it is usual and much more useful to describe the pulse in terms of a group of characteristic parameters. It is, of course, possible to define a large number of parameters describing such a pulse. In addition, these parameters can be defined in different ways. As an example, consider the parameter pulse duration. It could be defined in terms of the number of half-cycles above a

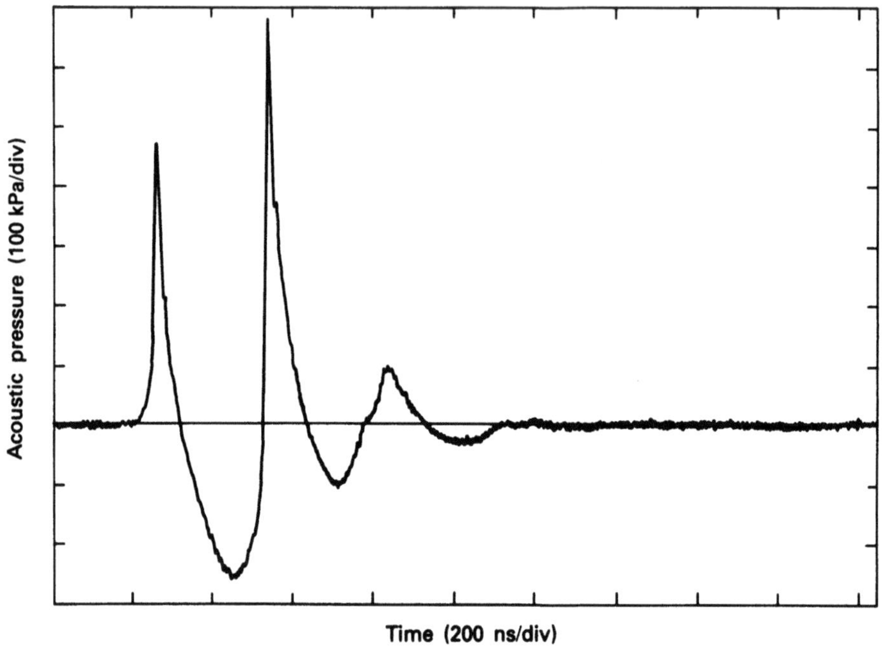

Figure 6.1. A typical acoustic waveform measured using a hydrophone.

certain amplitude, in terms of the time interval over which the integral of the square of the waveform corresponds to a certain fraction of the total integral, or in terms of many other methods. Each method would yield different values for the pulse duration. To avoid this situation and therefore to provide some degree of standardisation, various national and international bodies have produced relevant definitions of acoustic terms. The most important contributions are from the American Institute for Ultrasound in Medicine in collaboration with the National Electrical Manufacturers Association (AIUM/NEMA) [1], US Food and Drug Administration (FDA) [2] and International Electrotechnical Commission (IEC) [3,4].

Whilst these standards or draft standards contain a large number of definitions, these can be considered in a number of simple categories. This is done in the next sections which give simplified explanations of some of the most important acoustical parameters. A full description and a complete set of definitions of these and other parameters are given in the Glossary.

6.3 Pressure Parameters

By reference to Fig. 6.2, the most important parameters which can be derived directly from the pressure waveform are as follows:

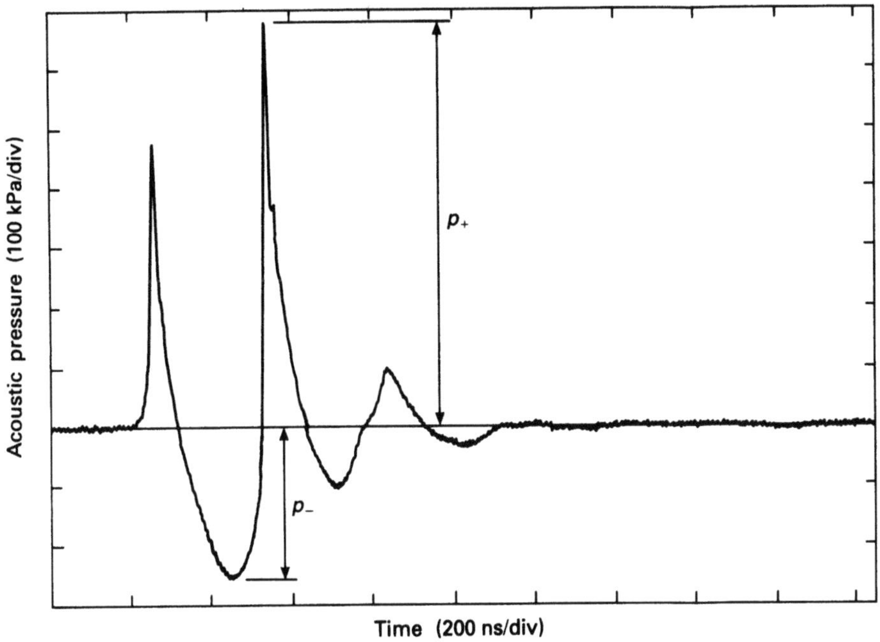

Figure 6.2. Acoustical pressure parameters p_+ and p_-.

Peak-Positive Acoustic Pressure, p_+ (or p_c):
Maximum positive instantaneous acoustic pressure in the pulse. p_+ is determined directly from the acoustic waveform knowing the sensitivity of the hydrophone measurement system.

Peak-Negative Acoustic Pressure, p_- (or p_r):
Maximum of the modulus of the negative instantaneous acoustic pressure in the pulse. p_- is also determined directly from the acoustic waveform.

Pulse-Pressure-Squared Integral, p_i:
Time integral of the square of the instantaneous acoustic pressure in the pulse, integrated over the whole of the pulse (see Fig. 6.3). p_i is determined from the integral of the square of the waveform, either numerically or digitally.

Pulse Duration, t_d:
Equal to 1.25 times the interval between the time when the time integral of the instantaneous acoustic pressure squared reaches 10% and 90% of its final value, the final value being the pulse-pressure-squared integral, p_i (see Fig. 6.3). Once p_i has been determined, t_d can be calculated from:

$$t_d = 1.25\,(t_2 - t_1)$$

where t_1 and t_2 are given by:

$$\int_0^{t_1} p^2 \mathrm{d}t = 0.1\,p_i \qquad \int_0^{t_2} p^2 \mathrm{d}t = 0.9\,p_i$$

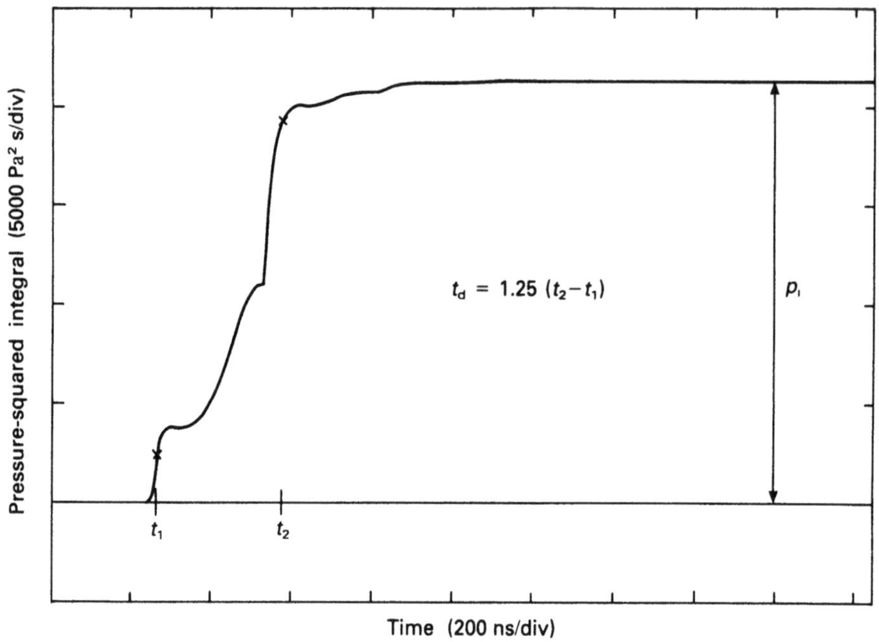

Figure 6.3. Time integral of the pulse-pressure-squared integral. p_i is the final value for the waveform shown in Fig. 6.2. t_d is the pulse duration determined from the pulse-pressure-squared integral.

Acoustic Working Frequency, f_{awf}:
The arithmetic mean of the frequencies f_1 and f_2 at which the amplitude of the acoustic pressure spectrum first becomes 3 dB lower than the peak amplitude. This is derived from a spectrum analysis of the pulse waveform, for instance by Fourier analysis (see Fig. 6.4).

Pulse Repetition Rate, *prr*:
The number of identical acoustic pulses per second. This is determined directly by determining the rate of arrival of identical pulses at the hydrophone.

The parameters described above refer to a particular waveform. In practice, measurements are made by moving the hydrophone to any point in the ultrasonic field, in which case each position will have a different waveform and therefore its own set of values for these parameters in order to describe the local field. Examination of the full definitions of the parameters (see Glossary and [3]) shows that many of them refer by default to the maximum value in the whole ultrasonic field. The term "spatial-peak" is then used to refer to the maximum value in an ultrasonic field. However, often they are used to describe the values of parameters at a specified position, say a particular measurement plane.

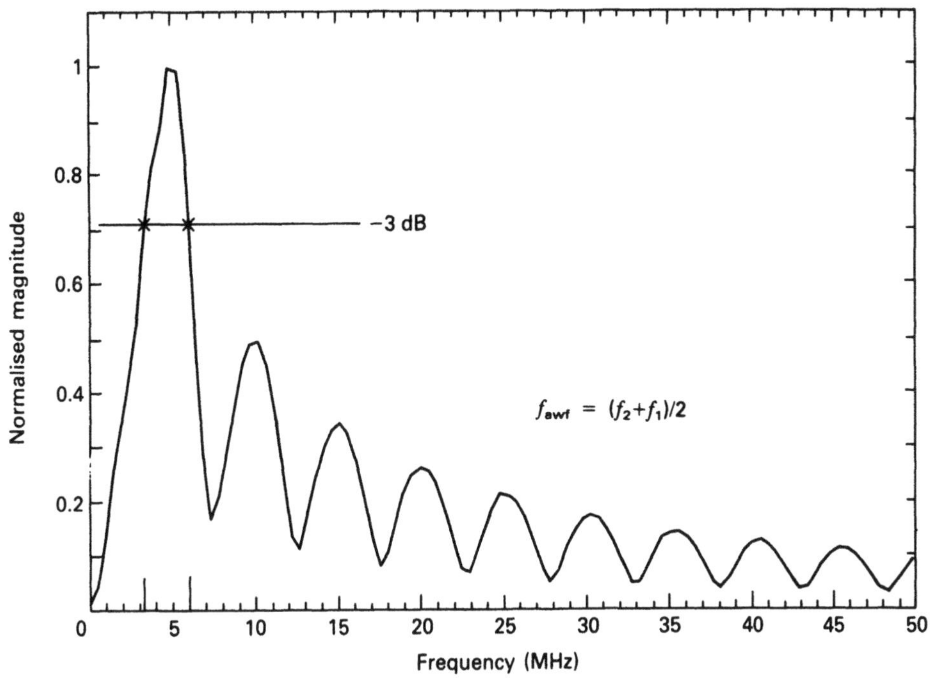

Figure 6.4. Spectrum analysis of an acoustic pulse waveform showing the derivation of the acoustic working frequency, f_{awf}.

6.4 Derivation of Intensity from Pressure Measurements

It has been traditional to specify acoustic fields in terms of intensity parameters. Instantaneous intensity can be derived from acoustic pressure measurements under assumptions of plane progressive waves by the relationship:

$$I = \frac{p^2}{\rho c}, \tag{6.1}$$

where ρ is the density of water and c is the speed of sound in water. At 20°C, the product ρc, the characteristic acoustic impedance of water, can be taken as 1.48×10^6 kg/m^2/s. Table 6.1 gives values at a range of temperatures.

Eq. (6.1) can be considered valid for most measurements of diagnostic ultrasonic fields but cannot strictly be used for measurements made close to the face of transducers. For instance, the ratio of the true axial intensity to the plane progressive wave value is 0.66 at a distance of 5 mm from the face of a circular plane transducer of diameter 30 mm. Hence, derivation of intensity close to the face of a therapy transducer will be in significant error. In practice, this does not mean that measurements cannot be made close to the face of a therapy transducer. It means that pressure measurements cannot easily be related to intensity. It will, however, be assumed here that intensity can be derived from pressure measurements.

Table 6.1. Values of density, ρ, speed of sound, c, and characteristic acoustic impedance of water at different temperatures, T

Temperature, T (°C)	Density, ρ (10^3 kg/m^3)	Speed of sound, c (m/s)	Characteristic acoustic impedance, ρc (10^6 kg/m^2/s)
16	0.9989	1469.4	1.4678
18	0.9986	1476.0	1.4739
20	0.9982	1482.3	1.4796
22	0.9978	1488.3	1.4850
24	0.9973	1494.0	1.4900
26	0.9968	1499.3	1.4945
28	0.9962	1504.4	1.4987
30	0.9957	1509.2	1.5026
32	0.9950	1513.6	1.5062
34	0.9944	1517.8	1.5094
36	0.9937	1521.8	1.5122
38	0.9930	1525.5	1.5148

6.5 Intensity Parameters

Fig. 6.5 shows the most important intensity parameters. These are:

Temporal-Peak Intensity, I_{tp}:
Maximum instantaneous intensity in the pulse. This is determined directly from p_+ or p_-, whichever has the largest magnitude, using Eq. (6.1).

Pulse-Intensity Integral, I_{pi}:
Time integral of the instantaneous intensity in the pulse, integrated over the whole pulse. I_{pi} is determined from the pulse-pressure-squared integral, using Eq. (6.1), by:

$$I_{pi} = \frac{p_i}{\rho c}$$

Pulse-Average Intensity, I_{pa}:
Ratio of the pulse-intensity integral to the pulse duration. This is directly derived from the determination of I_{pi} and t_d by:

$$I_{pa} = \frac{I_{pi}}{t_d}$$

Maximum Intensity, I_m:
Average intensity over the pulse half-cycle having the greatest temporal-peak intensity. This is determined from the instantaneous intensity by:

$$I_m = \int_{half\ cycle} I dt \,/\, half\text{-}cycle\ duration$$

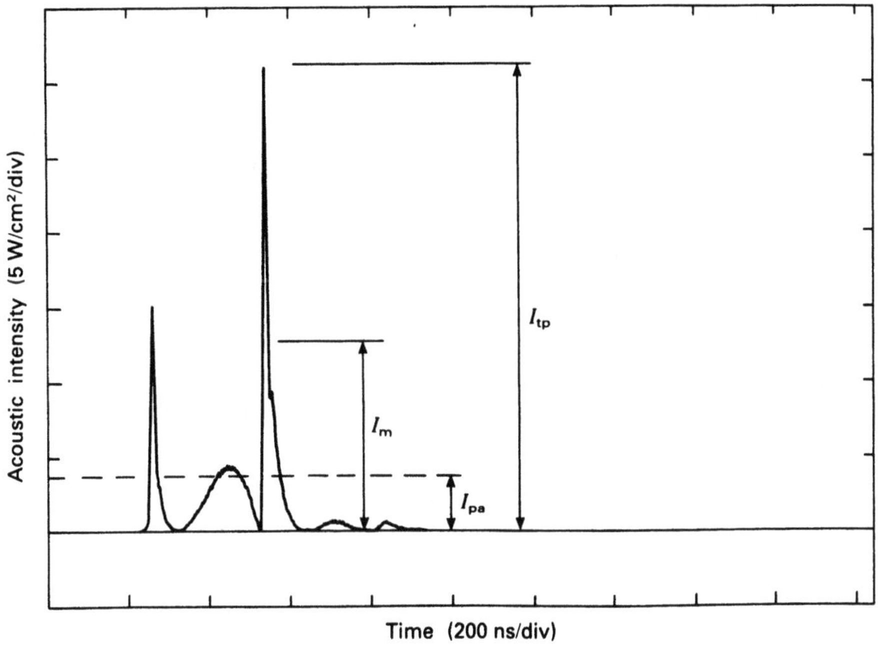

Figure 6.5. Intensity parameters for the same pulse as shown in Fig. 6.2.

Temporal-Average Intensity, I_{ta}:
Time-average of the instantaneous intensity. The time-average is taken over an integral number of acoustic repetition periods (1/*prr*) (see Fig. 6.6). If all pulses are identical then:

$$I_{ta} = I_{pi}\, prr$$

Whilst pulse-average and maximum intensities do not apply to continuous-wave fields, temporal-average intensity can apply in which case the time average is taken over an integral number of cycles of the waveform.

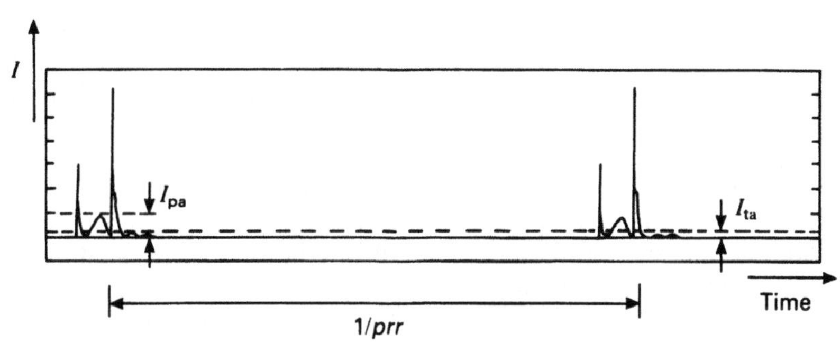

Figure 6.6. Intensity parameters I_{pa} and I_{ta} for a series of pulses at a pulse repetition rate *prr*.

6.6 Spatial-Peak Parameters

It is usual to specify the largest values of some of these acoustic parameters by using the term spatial-peak. In practice, spatial-peak can mean the largest value in the whole ultrasonic field or the largest in a specified plane. The most important parameters referred to in this way are:

Spatial-Peak Temporal-Peak Intensity, I_{sptp}:
Largest value of the temporal-peak intensity in a specified plane or in the whole ultrasonic field.

Spatial-Peak Pulse-Average Intensity, I_{sppa}:
Largest value of the pulse-average intensity in a specified plane or in the whole ultrasonic field.

Spatial-Peak Temporal-Average Intensity, I_{spta}:
Largest value of the temporal-average intensity in a specified plane or in the whole ultrasonic field.

For real-time scanning systems it is often appropriate to consider the acoustical parameters associated with a single acoustic scan line. In this case, the term "beam" is used rather than "spatial" – see [3] and Chapter 9.

6.7 Beam-Shape Information

By moving the hydrophone across the ultrasonic beam (in a direction normal to the propagation direction) the shape of the beam can be determined. This shape depends on the parameter that is being measured. An example of a typical beam shape is given in Fig. 6.7 where the pulse-pressure-squared integral has been plotted as a function of distance. Here, the hydrophone has been moved in a straight line passing through the point of maximum pulse-pressure-squared integral. It is possible to sample the beam by moving the hydrophone in other straight lines passing through the same maximum. For a cylindrically symmetrical beam these beam profiles would be identical. However, many beams are not cylindrically symmetrical, especially those produced by multi-element transducers such as phased and linear-phased array transducers. For such beams, it is often appropriate to determine the beam shape in two orthogonal directions. For real-time scanning systems, one of these directions is usually chosen to be parallel to the scan plane in order to define a reference direction. The beam-shape information in these two directions, parallel and perpendicular to the scan plane, is considered to be of most interest in specifying the focusing properties of the ultrasonic fields generated by medical ultrasonic diagnostic equipment.

For continuous-wave systems, the parameter plotted would be the mean square acoustic pressure. From a particular straight line sample, the beam width can be specified as:

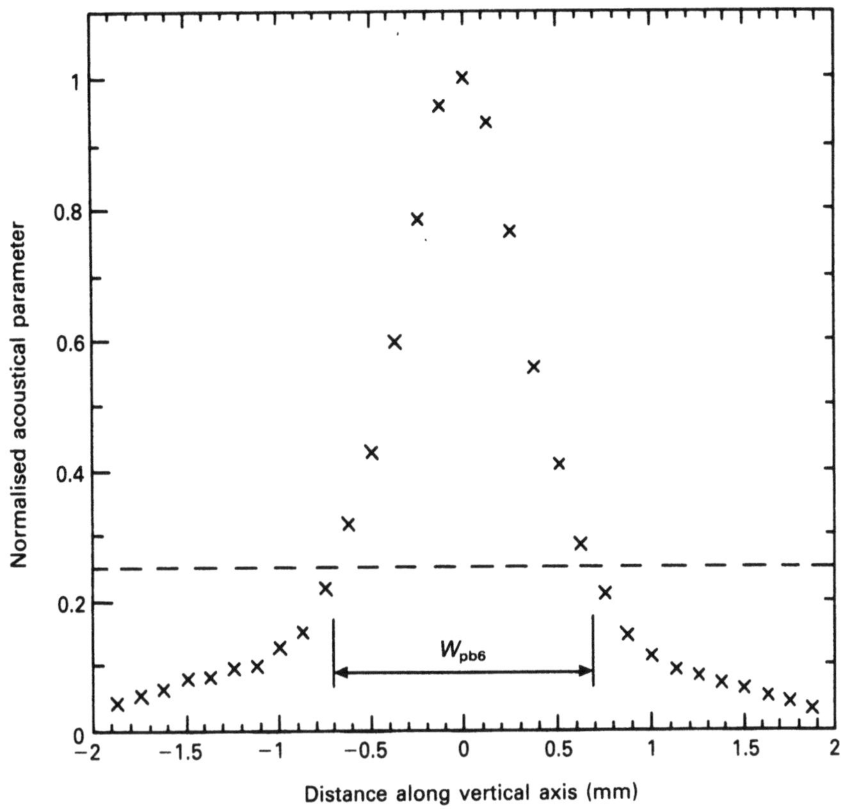

Figure 6.7. Typical beam profile in which the acoustical parameter plotted is the pulse-pressure-squared integral.

Pulse Beam-Width, w_{pb}:
Distance between two points, on a plane perpendicular to the ultrasound beam and in a specified direction passing through the point of the maximum pulse-pressure-squared integral, at which the pulse-pressure-squared integral is a specified fraction of the maximum value of the pulse-pressure-squared integral. Normally, the specified level is 0.25 and this is then referred to as the −6 dB pulse beam-width (w_{pb6}).

If a raster scan is made by taking parallel samples across the beam (see Chapter 7), equally spaced for simplicity, a beam area can be determined as:

Beam Area:
Area of the surface consisting of all points at which the pulse-pressure-squared integral is greater than a specified fraction of the maximum pulse-pressure-squared integral. Again, the specified level is 0.25 for the −6 dB beam area.

6.8 Spatial-Average Parameters

It is possible to average any one of the acoustical parameters over the beam area. This, of course, requires that the parameter being averaged is measured at each of the sample points. In practice, the only parameter averaged in this way is the temporal-average intensity, directly derivable from the pulse-pressure-squared integral.

Spatial-Average Temporal-Average Intensity; I_{sata}:
Average of the temporal-average intensity taken over the beam area. This is determined by integrating the temporal average intensity, I_{ta}, over the beam area and dividing by the beam area:

$$I_{sata} = \int_{\substack{beam\\area}} I_{ta} dA \, / \int_{\substack{beam\\area}} dA$$

For a raster scan consisting of a large number (>100) of equally spaced sample points, I_{sata} is then given by:

$$I_{sata} = \frac{\Sigma I_{ta} \text{ for all points at which } I_{ta} > 0.25\, I_{ta} \text{(maximum)}}{\text{Total number of points counted}}$$

I_{sata} can also be expressed in terms of the measured values of the pulse-pressure-squared integral p_i by:

$$I_{sata} = \frac{prr}{\rho c} \frac{\Sigma\, p_i \text{ for all points at which } p_i > 0.25\, p_i \text{ (max)}}{\text{Total number of points counted}}$$

In practice, the beam may be cylindrically symmetrical in which case the integration can be based on a diametrical sample of the beam rather than a raster sample.

6.9 Power Parameters

If the spatial integration of the beam is extended over all points which contribute significant signals then the power in the beam can be determined from:

$$\text{Power} = \int_{\substack{whole\\beam}} I_{ta}\, dA$$

6.10 Output Beam Intensity

Another parameter which is considered relevant to the assessment of heating by ultrasound is:

Output beam intensity, I_{ob}:
Power divided by the area of the beam at the output face of the transducer:

$$I_{ob} = \text{Power} / \text{Output beam area}$$

Normally, this parameter would be determined from the measurement of the total power using a radiation force balance (Chapter 5) although it is often easier and more convenient to determine power from a spatial integration of the pulse-pressure-squared integral as referred to in 6.9 and 7.4. To avoid errors in converting the pressure squared to intensity, the spatial integration would be undertaken at a distance from the transducer at least one or two times the diameter of the transducer face. However, the distance chosen would not be so great that errors from diffractive spreading or attenuation in the water path become significant. The advantage of the spatial integration technique is that it is possible to cope with real-time scanning systems which generate beams over a wide range of angles by treating each type of acoustic pulse separately. However, these procedures can be time consuming.

The output beam area (at the −6 dB level) would be determined from measurements at the face of the transducer although it is often possible to utilise the manufacturers specification for the transducer radiating area, at least as the basis for an initial estimate of I_{ob}.

6.11 Locating Maxima in Fields

The usual objective in making many measurements of the acoustic output of ultrasonic transducers is to determine the maximum value of a particular parameter. Unfortunately, the maximum value of different acoustical parameters occurs at different positions in the field. In general, the most important parameters of interest are:

Peak-positive acoustic pressure (p_+)
Peak-negative acoustic pressure (p_-)
Pulse-pressure-squared integral (p_i)

For a single acoustic pulse repeating at a rate *prr*, the position of the maximum value of p_i will also correspond to the position of maximum I_{spta}. Unless there are unusual changes in the pulse waveform, and therefore of the pulse duration, throughout the beam, the position of maximum p_i will also correspond to the position of maximum I_{sppa}. In the case of a scanning beam there may be significant differences between the positions of maximum p_i and of maximum I_{spta}, these are dealt with in Chapter 9.

In making measurements of an ultrasonic field, it is possible to draw a number of useful general conclusions concerning the location of the maximum values. These are:

1. In a plane perpendicular to the direction of propagation, the position of maximum p_+, p_- and p_i will be the same.
2. In the direction of propagation, the position of maximum p_+, p_- and p_i will nearly always be different.

The reason for the different locations in the propagation direction is the harmonic distortion of the waveform caused by nonlinear propagation of the ultrasound.

Fig. 6.8 shows the variation of p_+, p_- and p_i with distance from the face of the transducer. It is usually the case that p_+ peaks at the greatest distance, p_- at the smallest distance and p_i is maximum at a position between the two. It is for this reason that measurement in the plane containing p_i represents the best choice if only one set of measurements is to be made. Choosing the plane of maximum p_+ or p_- would incur a large error in the maximum values of p_- or p_+ respectively. If, however, the true maximum values for all parameters are required these measurements must be made at three different distances.

6.12 Concept of *in-situ* or Derated Levels

All acoustical output measurements of medical ultrasonic equipment are made with the field propagating in water. Whilst there are good arguments for using water as the *standard* measuring medium, it is not always representative of the real clinical situations. There are many approaches which could address this problem but the one most likely to be encountered is that proposed by the US Food and Drug Administration [2]. This assumes a constant attenuation coefficient model in which the intensity *in-situ*, I_{FDA}, is given by

$$I_{FDA} = I_w \exp(-0.069 fz) \tag{6.2}$$

where I_w is the measured intensity in water, f is the ultrasonic frequency in MHz and z is the distance in cm. The actual value of the attenuation coefficient used by FDA is 0.0345 Np/MHz/cm or 0.3 dB/MHz/cm. This particular value was chosen to be representative of typical low-loss tissue, so that the resultant *in-situ* levels should be high, thereby being more relevant to safety considerations.

The significance of this attenuation can be seen from Fig. 6.9 which gives the variation of $\exp(-0.069 fz)$ with distance for four typical ultrasonic frequencies. The overall effect increases rapidly with distance from the transducer face, especially for the higher frequencies.

It is possible to apply Eq. (6.2) to determine the *in-situ* levels from any water measurements provided the distance, z, between the transducer output face and the point of interest is known and also that the frequency is known. However, the original purpose of the *in-situ* model was for regulation and this required that the position of maximum *in-situ* exposure is found. Often the

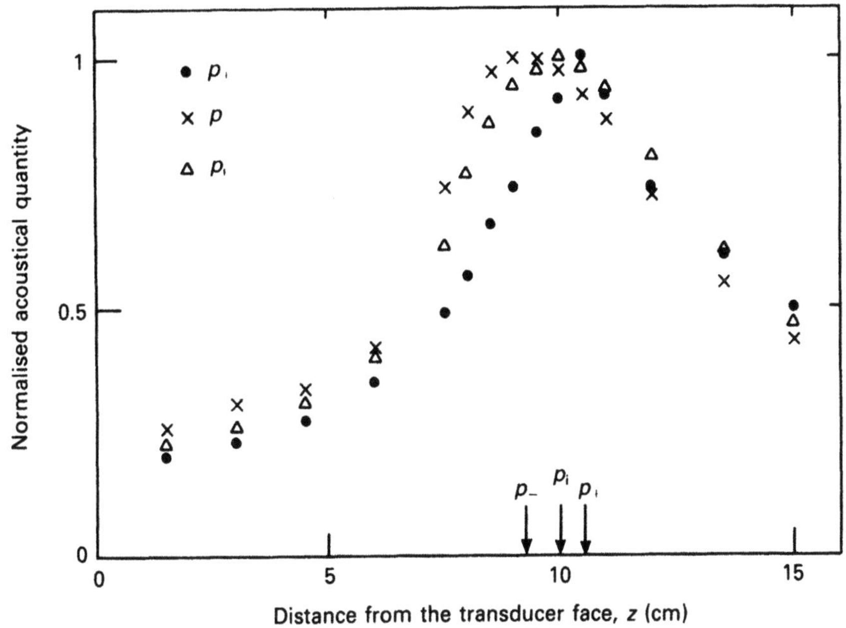

Figure 6.8. Typical axial plot of p_+, p_- and p_i.

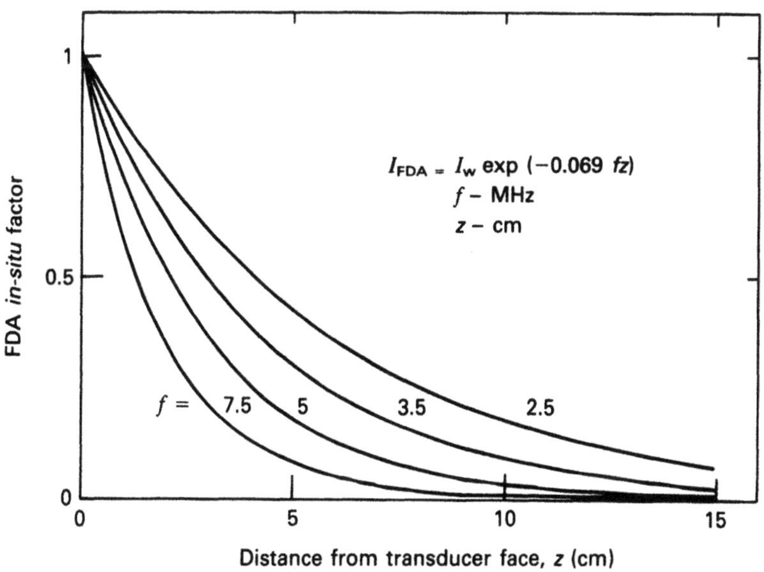

Figure 6.9. FDA *in-situ* or derated factor as a function of distance, z, at different frequencies, f.

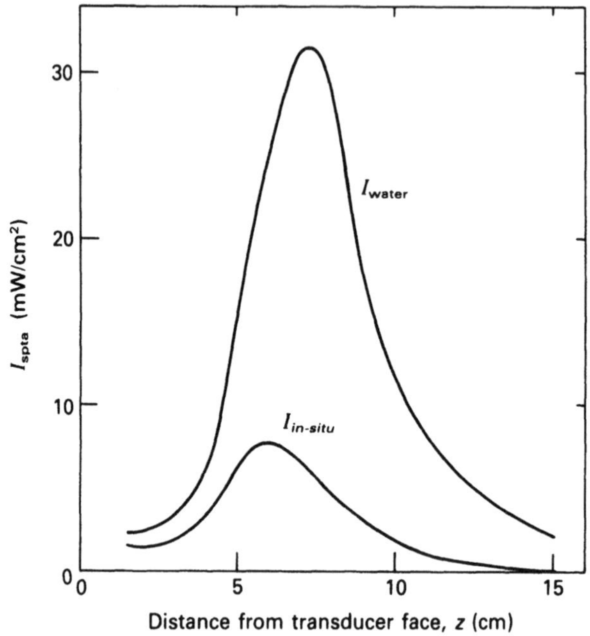

Figure 6.10. Axial variation of I_{spta} measured in water and the derived FDA *in-situ* levels using the FDA *in-situ* factor.

maximum *in-situ* intensity occurs at a different distance from the transducer face compared with that of the maximum water intensity. This can be seen in Fig. 6.10 where the axial variation for a typical focused field has been plotted in water and then the *in-situ* values calculated at each point. The maximum *in-situ* level appears slightly closer to the face of the transducer than for the case of water measurements.

Current policy of declaring the acoustic output of medical ultrasonic equipment solely in terms of *in-situ* levels using the attenuation coefficient of 0.3 dB/MHz/cm can be misleading. It is often not made clear that the figures refer to *in-situ* levels. The major problem with the specification of only *in-situ* levels is that it is not possible to use them as the basis for the prediction of exposure levels in other critical clinical situations. This is especially true for obstetric examination where overall attenuation between the transducer face and the foetus may be quite low in comparison to the attenuation calculated using the 0.3 dB/MHz cm. This is because the propagation path may contain a significant proportion of low attenuating fluids such as amniotic fluid or urine. The limitations of this approach have been realised and for this reason the term *in-situ* is often replaced by "derated" in an attempt to emphasise that the assumptions used are such that the levels predicted do not closely relate to true *in-vivo* levels. Both terms are taken to have the same meaning as defined by Eq. (6.2).

The estimation of *in-situ* exposure will gain importance over the next few years as different methods are examined in order to find those which give the most accurate method of predicting levels likely to arise in tissue. Methods may depend on the clinical examination being undertaken and in future standard layered models may be established for crucial clinical scanning situations. It is probable that the US National Council on Radiation Protection and Measurements [5] will be publishing information on recommendations for simple models. The International Electrotechnical Commission (IEC) will also be addressing this issue in the future.

6.13 References

1. AIUM/NEMA. Safety standard for diagnostic ultrasound equipment. Publication UL 1-1981, National Electrical Manufacturers Association, Washington DC 20037, USA, 1981
2. FDA. Guide for measuring and reporting acoustic output of diagnostic ultrasound medical devices. Document 510(k), US Department of Health and Human Services, Food and Drug Administration, HFZ-132, Rockville, MD 20857, USA, 1985
3. IEC. Measurement and characterisation of ultrasonic fields using hydrophones in the frequency range 0.5 MHz to 15 MHz. To be published as an IEC Standard, Document 87(CO)6, International Electrotechnical Commission, Geneva, Switzerland, 1988
4. IEC. Requirements for the declaration of the acoustic output of medical diagnostic ultrasonic equipment. Draft International Standard, Document 87(CO)11, International Electrotechnical Commission, Geneva, Switzerland, 1990
5. NCRP. Exposure criteria for medical diagnostic ultrasound. Part 1. Criteria based on thermal mechanisms. Draft Report of Scientific Committee No. 66, National Council on Radiation Protection and Measurements, Bethesda, MD 20814, USA, 1990

7

Measurements on a Specific Acoustic Pulse Part 2: Measurements

D.R. Bacon

7.1 Introduction

This chapter describes the procedures used and the difficulties encountered when trying to measure the acoustic parameters mentioned in Chapter 6. Section 7.2 deals with the measuring instrumentation, 7.3 and 7.4 describe the measurement procedure, 7.5 deals with problems specific to calibrating continuous-wave systems and finally 7.6 discusses measurement accuracy. The whole of this chapter concentrates on measurements on a piece of equipment in a single excitation mode; Chapters 8 and 9 deal with features specific to real-time scanning devices.

7.2 The Measurement System

It is assumed here that the basic measurement system consists of a hydrophone, connected to a voltage measuring device (an oscilloscope or digitiser), probably using a preamplifier between the two. Some of the characteristics of hydrophones have already been discussed in Chapter 4, but here their significance in measurement is described.

First, however, the requirements for the test tank and the transducer mounting arrangements will be discussed; the hydrophone mount has already been described in 4.3.2. Three possible tank geometries are shown in Fig. 7.1. Ideally, the transducer face should be immersed in water, to eliminate any coupling problems due to variations in the contact between the transducer and any membrane. However, some transducers are not completely waterproof, so that a membrane has to be used in some cases. The transducer mount does not need to provide any translation capability if the hydrophone mount provides this (see 4.3.2), but it does need to allow orientation of the transducer by

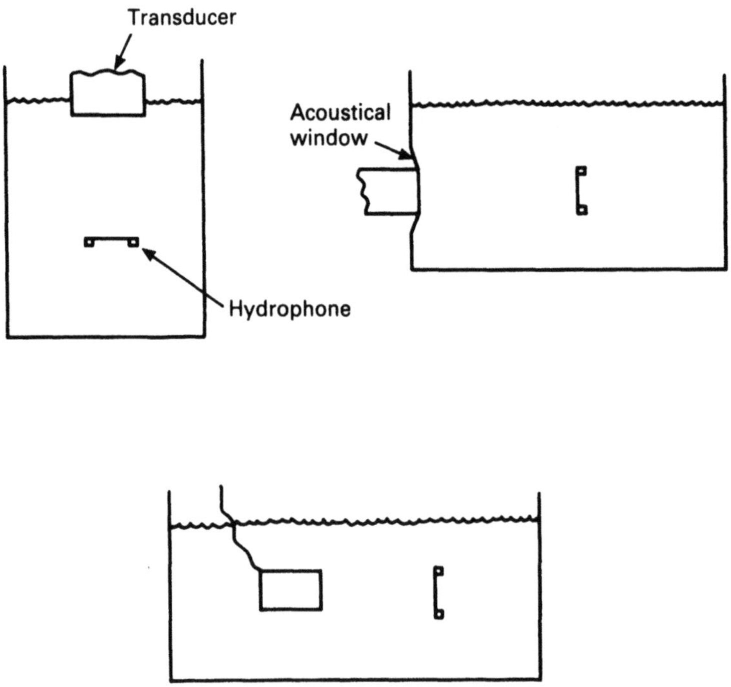

Figure 7.1. Schematic diagrams of three mounting arrangements for test tanks.

rotation about two orthogonal axes in a plane parallel to its face (see 7.3.1). Overall, the system should allow relative movement of the hydrophone and transducer over three orthogonal axes. The arrangements where the ultrasound beam propagates horizontally are particularly convenient for systems where a large number of degrees of freedom are provided (such as automatic scanning systems, see [1]). The arrangement that incorporates a vertical ultrasound beam allows the rapid exchange of transducers, although the hydrophone is more difficult to manipulate [2,3].

The design of the transducer clamp is an important feature of any system, since the clamp has to cope with a large range of transducer sizes and geometries, whilst permitting the orientation to be adjusted. A laboratory retort-stand clamp is one of the simplest designs, which can be quite effective. Although it needs to be custom-made, a simple vice-like arrangement can also be used, especially if it can be designed to cope with a range of diameters up to 50 mm. A long, open-ended vice is also particularly useful for holding linear array transducers.

7.2.1 System Bandwidth

Hydrophones have a certain bandwidth (usually between 20 and 100 MHz) over which they have a significant sensitivity and they do not respond to frequencies outside this range. Furthermore, they usually have a resonance in their response (see 4.3.7), where the sensitivity increases by a factor of about 1.5–2.0 for polyvinylidene fluoride (pvdf) devices. These limitations are particularly noticeable when measuring pulses containing shock fronts produced by nonlinear propagation. The limited bandwidth can give rise to a smoothing of the steep rising edge of the shock front, leading to an underestimate of the peak-positive pressure (Fig. 7.2a). On the other hand, the effect of the hydrophone resonance is to cause overshoot in the response just after the shock front has arrived, leading to an overestimate of the pressure (Fig. 7.2b). An ideal situation would be one where the bandwidth is as wide as possible, but with minimal effects due to resonances. To achieve this, it is necessary to consider the behaviour of the whole system, because electrical resonances in the hydrophone cable (see 4.3.7) and bandwidth limitations of the preamplifier and electrical measuring instrument can also be important. It is sometimes possible to adjust the bandwidth of the system by changing the oscilloscope/digitiser bandwidth or by adjusting the preamplifier. If this is possible, then one should aim to achieve an overall frequency response which is smooth, i.e. with little increase in response with frequency and a gradual decrease around the cut-off frequency. To achieve this, it may be necessary to sacrifice overall bandwidth – for example a system with a 25 MHz bandwidth may perform better than one with a 40 MHz bandwidth but which includes a sharp resonance. A comparison of different systems has been given by Smith [4].

Once the optimum system has been achieved, it may be possible to obtain an improved estimate of the true peak-pressure by fitting a curve to the smooth part of the waveform and then extrapolating to find its value at the position of the shock front (Fig. 7.3). The details of one such method are given by Preston et al. [1].

7.2.2 Spatial Resolution

Hydrophones have a limited spatial resolution and this is largely determined by the size of their active element. Membrane hydrophones, however, tend to have poorer resolution than this at frequencies below about 2 MHz. For a membrane hydrophone, the so-called effective radius is given approximately by:

$$a \approx \left(a_g^2 + \frac{1}{4f^2}\right)^{\frac{1}{2}} \tag{7.1}$$

where f is the frequency in MHz, a_g is the geometrical radius of the hydrophone and a is the effective radius, both expressed in millimetres. Fig. 4.6 gives typical plots of a against frequency for membrane hydrophones.

There are two ways of determining the correct size of hydrophone to use in a particular application. The first is to measure the beam-width of the field and compare this with the effective dimensions of the hydrophone. This method is satisfactory if the effective diameter ($2a$) is less than one third of the −6 dB beam-width (Fig. 7.4, see [5]). If this condition is met, then the error in the measurement of peak pressure should normally be less than 3%. Above this size of hydrophone, the error can increase rapidly, particularly if nonlinear propagation is present. For instance, a hydrophone with a diameter equal to

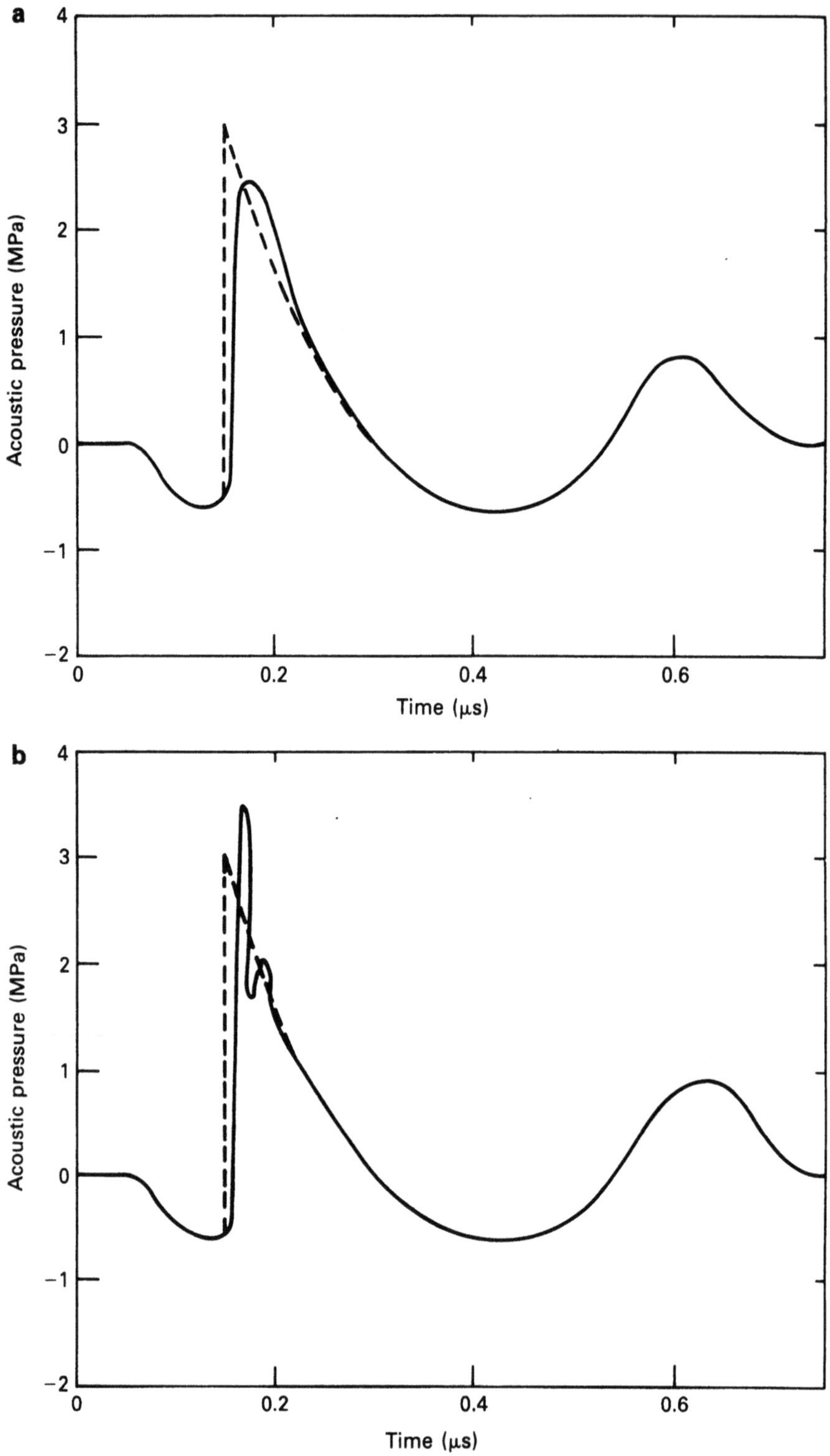

Figure 7.2. Waveforms to illustrate the effect of the hydrophone bandwidth on the measured waveform. *Dashed line*, true acoustic waveform; *continuous line*, measured acoustic waveform. **a** Hydrophone with a limited bandwidth and little resonance in its response; **b** hydrophone with a resonance in its frequency response.

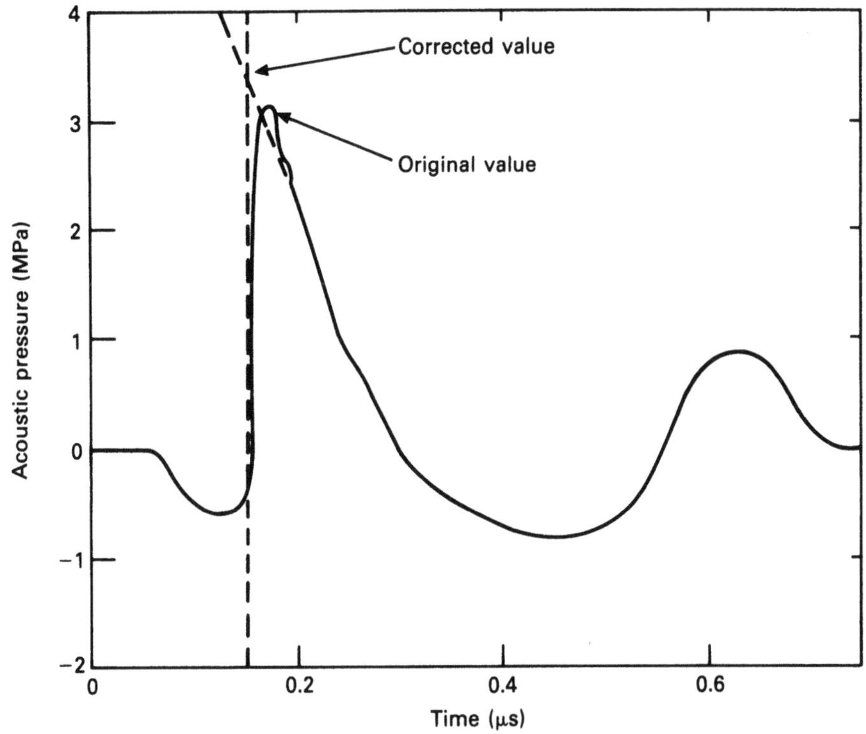

Figure 7.3. Illustration of the method for correcting the measured signal for the effect of the hydrophone frequency response. *Continuous line*, measured waveform; *dashed vertical line*, position of the start of the shock front; *dashed angled line*, extrapolation of the measured waveform. The intersection between the two dashed lines gives the estimated value of the peak-positive acoustic-pressure.

half of the −6 dB beam-width can introduce errors in the measurement of pressure that range from 10% to 20% [5]. This first method of determining the hydrophone size assumes that one has some *a priori* knowledge of the field and it may therefore be somewhat cumbersome to apply in practice. The International Electrotechnical Commission (IEC) [6] gives another method for determining the required hydrophone size which requires a less detailed knowledge of the field. In this case the effective radius a must satisfy the following condition:

$$a \leqslant \frac{\lambda}{8}\left(1 + \frac{l^2}{a_1^2}\right)^{\frac{1}{2}} \tag{7.2}$$

where l is the distance from the source and a_1 is the transducer radius. If this formula is adhered to, then the measurement error should be less than approximately 5% provided that nonlinear propagation is not present. Fig. 7.5 gives the variation of the maximum permitted hydrophone size with distance, according to Eq. (7.2).

For a membrane hydrophone, the effective radius of the hydrophone is given by Eq. (7.1) in terms of the geometric radius. It is, therefore, possible to combine Eqs. (7.1) and (7.2) to obtain the following criterion for the geometric radius:

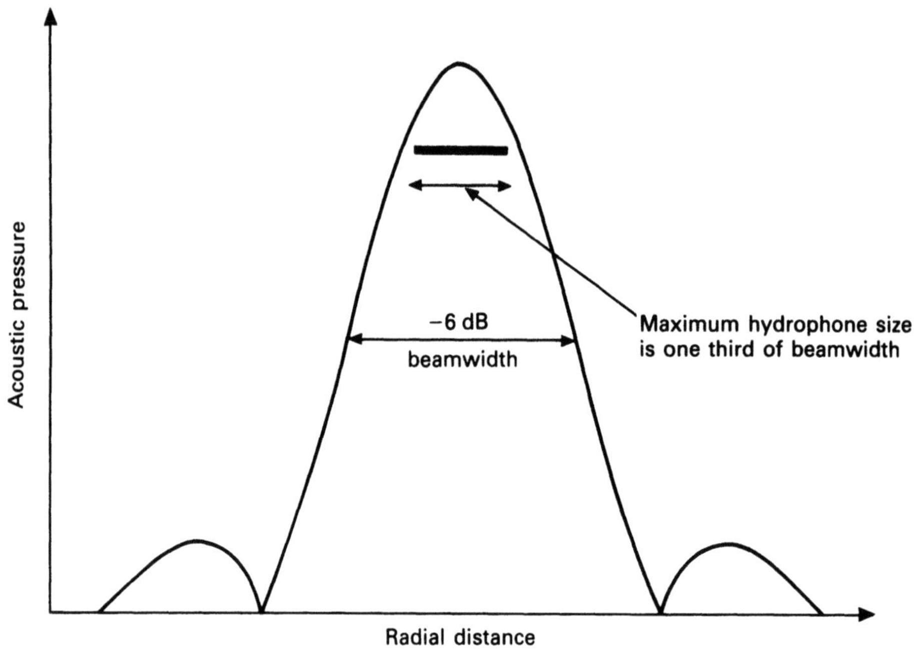

Figure 7.4. Illustration of the method for choosing the correct hydrophone size to avoid errors due to spatial averaging over the sensitive element. The effective hydrophone diameter should be less than one third of the −6 dB beamwidth of the acoustic field.

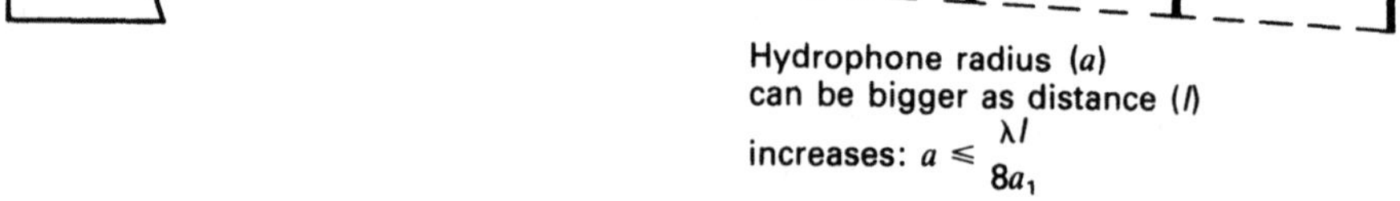

Figure 7.5. Illustration of the IEC method for choosing the correct hydrophone size, so as to avoid errors due to spatial averaging over the sensitive element. The maximum effective hydrophone size permitted by Eq. (7.2) is shown as a function of distance from the transducer.

$$a_g \leq \frac{\lambda}{8}\left(\frac{l^2}{a_1^2} - 2.5^2\right)^{\frac{1}{2}} \tag{7.3}$$

For large distances from the transducer, this relationship is similar to Eq. (7.2), with a replaced by a_g, but for small distances it shows that no membrane hydrophone will satisfy the IEC criterion. This limiting distance is equal to 2.5 a_1 (see Fig. 7.6).

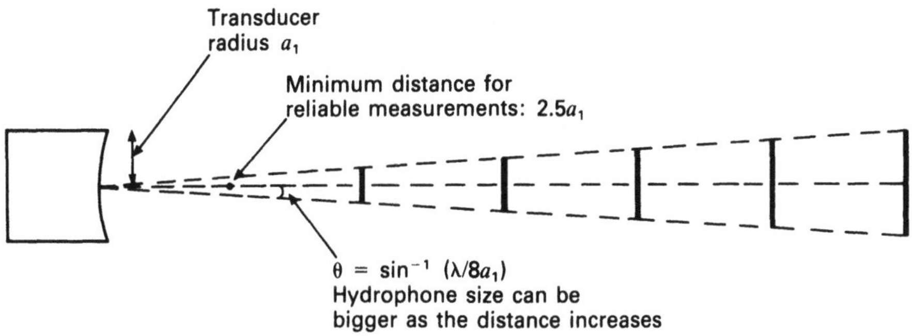

Figure 7.6. Illustration of the IEC method for choosing the correct hydrophone size, as applied to membrane hydrophones. The maximum geometric hydrophone size permitted by Eq. (7.3) is shown as a function of distance from the transducer.

Often it is not possible to use the optimum size of hydrophone, because it may not be available. It is also important to consider the fact that hydrophones with smaller active elements are less sensitive. To achieve adequate capture of the waveform a reasonable signal-to-noise ratio is required and so a compromise between spatial resolution and hydrophone sensitivity has to be reached. More information on spatial averaging is given by Beissner [7], IEC [6], Preston et al. [1] and Smith [5].

7.2.3 Digital Acquisition Systems

To make the best use of the information contained in a hydrophone signal, it is necessary to use a digitiser. This converts the analogue waveform to an array of data points, each of which corresponds to the voltage at a specific time (with a constant time increment between points). Each array variable can only take one of a limited range of values (there are 256 possible values for an 8-bit system). As a result, the resolution in both voltage and time is limited. Furthermore, all of the required points may not be captured by the digitiser in a single waveform, so that repetitive sampling of the waveform may be needed. This can give rise to difficulties. As an example, if there is 50 Hz interference on the signal then each successive waveform will start from a different voltage level and the digitised waveform will suffer from what appears to be high frequency noise. Another feature of using digitisers is that the ground level often needs to be carefully defined, as otherwise large errors can occur when the relevant acoustic parameter is calculated from the waveform array.

The limited time resolution may mean that a very sharp peak in the waveform may not be captured reliably (see Fig. 7.7a). If the time-resolution is poor it may be useful to interpolate between two waveform array points to gain better precision for the level at intermediate times. However, as illustrated by Fig. 7.7a, this may not improve the accuracy if the peak value of the waveform has not been detected. When calculating the results of a measurement, any

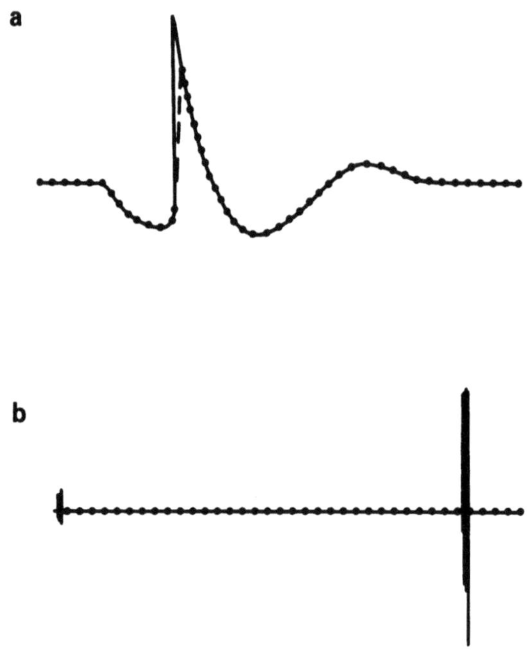

Figure 7.7. Illustration of the effect of a limited sampling rate for digital data acquisition. *Continuous line*, true waveform; *circles*, acquired data; *dashed line*, apparent waveform. For a typical waveform containing a shock front (**a**), most of the waveform may be captured accurately, but the peak pressure may be under-estimated. For an extreme case of under-sampling (**b**), the waveform may not be detected at all.

time integrals in the basic formulae (Chapter 6) need to be replaced by sums and the result of the sums multiplied by the sample interval of the digitiser.

A further consequence of the limited time resolution of a digitiser can be experienced when setting up the system for measurements. At this point in the procedure, a relatively slow time-base is usually employed and, as the number of samples per division is held constant, the sample interval can be relatively large. This can lead to severe undersampling of the waveform so that in extreme cases the ultrasound pulse can disappear from the screen altogether (Fig. 7.7b). Some digital oscilloscopes overcome this difficulty by providing an "envelope mode" of operation which displays the envelope of the waveform being captured.

7.2.4 Analogue Acquisition Systems

Ultrasound beam-characterisation systems that incorporate a digitiser are undoubtedly the most flexible, but there is nevertheless a role for analogue systems. In general these are simpler and more portable than their digital counterparts, largely because most digitisers are used in conjunction with a computer. As most laboratories will possess a suitable oscilloscope, an analogue acquisition system can make the best use of pre-existing equipment and be a more economical alternative.

If a hydrophone is used to detect the ultrasonic field, it can be connected directly to an oscilloscope of suitable bandwidth and sensitivity. As with a digital system, the oscilloscope needs to have a delayed time-base with a retrigger facility after the delay. Once the system has been triggered (see 9.2.4 and 9.2.5) and the peak waveform found, then it is straightforward to determine the peak-positive and peak-negative pressures (see 7.3.3). It is more difficult to obtain temporal average properties in the field, but the pulse-pressure-squared integral (p_i) can be estimated from the amplitude and duration of each half-cycle (see 7.3.3). To perform the necessary time-integrals automatically, Martin [2] has connected a radio-frequency power meter to the amplified output signal from the hydrophone. This gives a real-time indication of the temporal-average intensity and permits one to find the maximum value of this parameter in the field. As the power meter integrates the entire signal from the hydrophone, care has to be taken to eliminate spurious contributions. Multiple reflections of the ultrasonic wave from within the test-tank should be avoided, or else their effect estimated and corrected for. The electrical interference caused by pick-up of the transducer drive signal should be reduced by using a well-shielded hydrophone (e.g. the bilaminar-shielded membrane type). Furthermore, the effect of any residual interference should be estimated by placing an acoustical absorber in front of the hydrophone to eliminate the acoustic signal and thereby measuring the remaining noise signal. The effect of environmental electromagnetic radiation and noise in the system should also be checked and subtracted out. To minimise these effects, the hydrophone should be as sensitive as possible; to achieve this, a hydrophone with an active element 1 mm in diameter is normally used. With such a system, a typical signal-to-noise ratio of 50 : 1 has been achieved [2].

Portable ultrasonic power balances come under the category of analogue instruments and can be used to measure the output power of a range of medical devices (Chapter 5). They can also be used in identifying those modes of operation of a scanner that are likely to give the highest acoustic field levels, since such modes are likely also to give the highest output powers. Before performing such measurements, however, one needs to be sure that the whole output beam will be intersected by the balance, at normal incidence to the target. Frequently, these conditions cannot be met and so it is difficult to determine the true acoustic power (see 5.5.5).

7.3 Spatial-Peak Measurements

This section considers the measurement of the maximum value of an acoustic parameter in the ultrasonic field. Before starting measurements, it is advisable to perform some basic checks, such as verifying that the hydrophone amplifier is connected to a 50 Ω load if it needs one, and that the transducer is waterproofed with its surface free from air bubbles.

7.3.1 Finding the Acoustic Axis

Two procedures for finding the acoustic axis are outlined here, based on IEC 87(CO)6 [6]. At all times, the hydrophone is assumed to have been rotated so that it gives its maximum signal. The first procedure involves moving the hydrophone in a plane close to the transducer (e.g. 50 mm away) and noting the position where the signal is maximum. The hydrophone is then moved to a plane which is, for instance, 150 mm away and the process is repeated. From this information, the required rotation of the transducer is calculated such that the line joining the two positions of maximum pressure is parallel to the main axis of the positioning system. This procedure requires the angular position of the transducer to be calibrated, which may not always be the case.

A second possible procedure again involves maximising the hydrophone signal at two different transducer–hydrophone separations. This time, the source transducer is moved to plane A (at a significant distance from the hydrophone) and rotated to maximise the signal from the hydrophone (Fig. 7.8a). It is then moved to plane B (closer to the hydrophone) along the main positioning axis, where the signal is again maximised, this time by moving the hydrophone (Fig. 7.8b). This process is then repeated several times, until no remaximisation of the signal is required on moving from one plane to the other (Fig. 7.8c–e). The number of iterations required to achieve satisfactory alignment will be fewer if the distance between the two planes is kept as large as possible.

7.3.2 Finding the Measurement Plane

The first consideration here is to decide which acoustical parameter needs to be maximised in order to find the measurement plane, since the peaks of different parameters usually occur in different planes (see 6.11). Often the pulse-pressure-squared integral, p_i, is to be maximised, in which case an automatic measuring system is required, or at least one with a real-time readout of p_i (see 7.2.4). If the positioning system has been aligned as described in 7.3.1, then a simple scan along the main (z) axis will give the required measurement plane. Otherwise, the signal will have to be remaximised in the x and y directions for each z position, which will be very time-consuming.

7.3.3 Calculation of Acoustic Parameters

Having obtained a satisfactory waveform, the first step is to determine its acoustic working frequency. This can be estimated from the duration of a complete cycle of the pulse, but it is better to obtain a Fourier Transform of the signal as described in 6.3.

Having found the acoustic working frequency, the system sensitivity at this frequency is determined. This depends on the product of the hydrophone sensitivity, the amplifier gain and loading correction and any calibration factor of the digitiser or oscilloscope. The loading correction, L, takes account of the reduction in the hydrophone output voltage that occurs when the

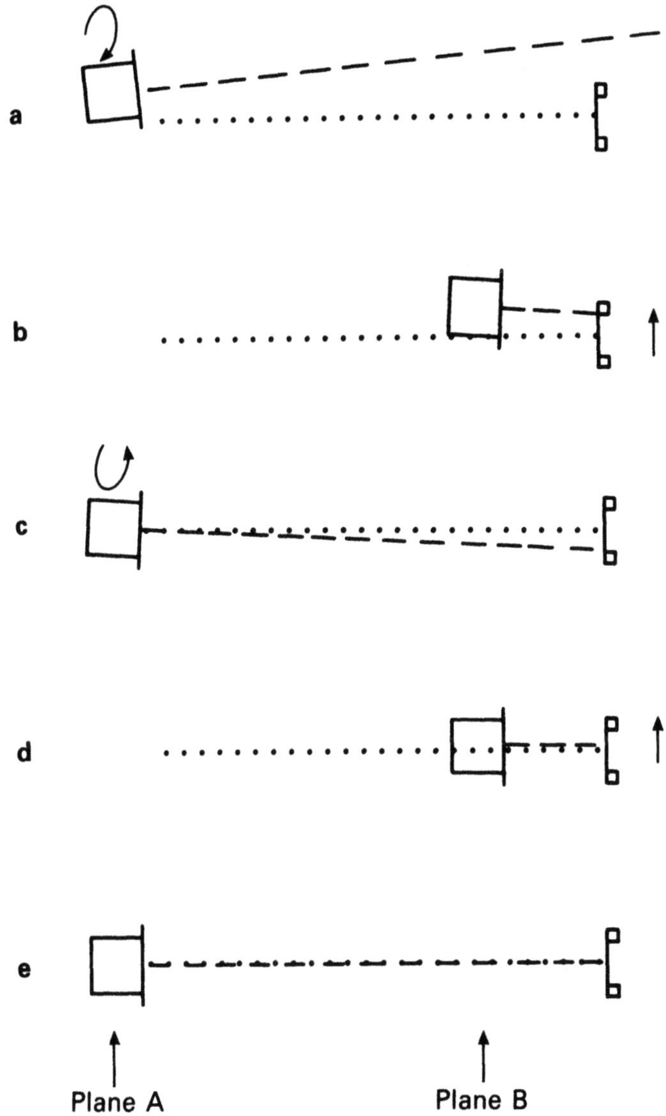

Figure 7.8. One method of aligning the acoustic axes of the source transducer (*on the left*) and the hydrophone (*on the right*). In a typical sequence: **a** the transducer is placed at a distance (in plane A) and rotated to give the maximum signal from the hydrophone; **b** the transducer is then placed closer to the hydrophone (in plane B), which is then moved to intersect the centre of the beam. This process is repeated (**c–e**) until the two axes are aligned.

hydrophone is connected to the input of the amplifier or oscilloscope. A complete method for calculating this should be given with the hydrophone calibration data, but an approximate formula (cf. Eq. (4.2)) is:

$$L = \frac{C}{C + C_{el}} \tag{7.4}$$

where C is the hydrophone capacitance and C_{el} is the capacitance of the amplifier or oscilloscope to which the hydrophone is connected. If any extra cable is used in making the connection to the amplifier or oscilloscope, then its capacitance must also be included in C_{el}. From Eq. (7.4), it can be seen that the loading factor is always less than unity.

The acoustic pressure, p, is calculated by dividing the measured voltage, V, by the system sensitivity, M_l, as follows:

$$p = V/M_l \tag{7.5}$$

To calculate the spatial-peak temporal-peak intensity I_{sptp}, the maximum value of p^2 needs to be obtained. This may occur when the peak pressure is either positive or negative, although it usually occurs for a positive value of the acoustic pressure. As the maximum peak-positive and peak-negative acoustic pressures usually occur at different places in the field, any maximisation routine should record both of these parameters. Similar considerations apply in finding the position where the maximum intensity I_m is maximum and for calculating I_m.

If the digitiser has a waveform processing facility, then it may be possible to use this to calculate various parameters. The energy function, which corresponds to the time integral of V^2, may be converted to give the pulse-pressure-squared integral, p_i, by dividing by M_l:

$$p_i = \frac{\int V^2 \mathrm{d}t}{M_l^2} \tag{7.6}$$

Similarly, p_i can be obtained from the root-mean-square voltage of the waveform, V_{rms}, as follows:

$$p_i = \frac{V_{rms}^2 t_w}{M_l^2} \tag{7.7}$$

where t_w is the time window (in seconds) over which the waveform was acquired. If an analogue oscilloscope is used, then it is possible to estimate the integral in Eq. (7.6) by noting the amplitudes V_n and durations t_n of each half-cycle. The integral can then be expressed as:

$$\int V^2 \mathrm{d}t \approx \sum \gamma_n V_n^2 t_n \tag{7.8}$$

where $\gamma_n = 0.5$ when the half-cycle appears to be sinusoidal and $\gamma_n = 0.33$ when the half-cycle appears to be triangular, due to the presence of a shock front. A similar approach can also be used to estimate the pulse duration. If a power meter is used to detect the hydrophone signal, then the integral in Eq. (7.6) can be performed automatically, provided that an electronic gate is used to select the particular pulse of interest [2]. In this case, p_i is given by:

$$p_i = \frac{P_{el} R}{M_l^2 prr} \tag{7.9}$$

where R is the input impedance of the power meter, P_{el} is the measured electrical power and prr is the pulse repetition rate (i.e. the repetition frequency for the electronic gate). For this case, care should be taken to ensure that the

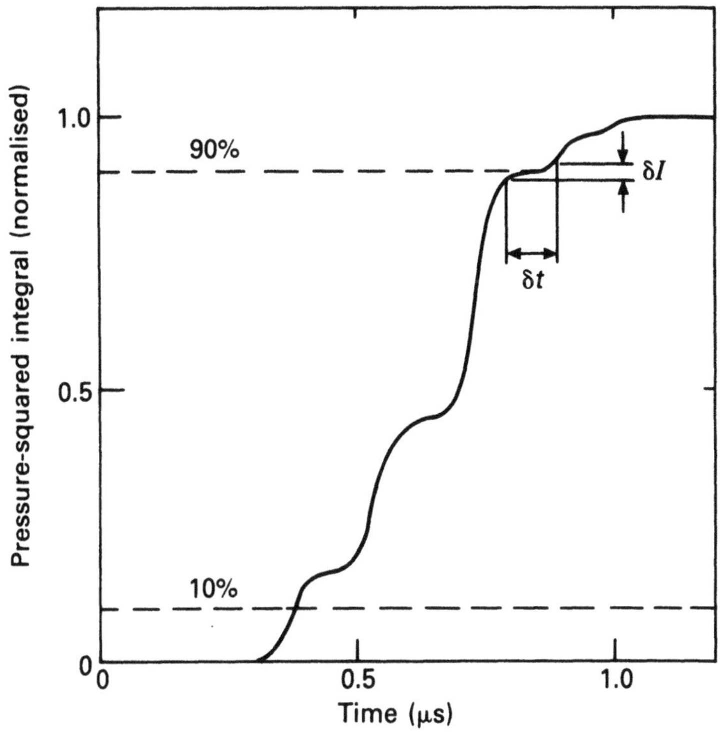

Figure 7.9. Diagram to illustrate how the pulse duration can have a large random variation. *Continuous line*, measured normalised pressure-squared integral as a function of time; *dashed lines*, 10% and 90% values of the pulse-pressure-squared integral.

system gain M_l is determined when operating into the input impedance (R) of the power meter.

When measuring the pulse duration (see 6.3), a repeated measurement may give very different results. An explanation for this can be given with reference to Fig. 7.9. Here, the integral of p^2 is plotted against time and shows a number of steps, where the position of the horizontal portions correspond to zero-crossings in the waveform. If one of these horizontal portions corresponds to the position where the integral is 10% or 90% of the maximum, then a small change in the integral (δI) will give rise to a large change in the value for the pulse-duration (δt). In this situation the result will still be valid, but an appropriately large random uncertainty should be assigned to take account of the problem.

7.4 Spatial-Average Measurements

To perform measurements of spatial-average parameters it is necessary to scan the hydrophone over the acoustic field (see 6.7). This section deals with some of the points to note when setting up such a scan. It will be assumed that an automatic system (e.g. digitiser and computer) is being used. Such a system is normally required for this work, due to the large number of measurements that have to be recorded.

7.4.1 Deciding on the Scan Parameters

A raster scan is the easiest type of scan to visualise, since it involves moving over equally spaced points within a rectangular region of the measurement plane. Since all of the points in this region are covered it gives the most complete information, but it takes rather a long time to accomplish. In order to obtain a reasonable estimate for the spatial-average temporal-average intensity using this type of scan it is necessary for there to be 25–100 points within the −6 dB area of the beam.

A one-dimensional scan, referred to here as a "beam-plot", takes less time but offers less complete information. If the beam is cylindrically symmetrical then good results can be obtained with less than 20 scan points. The scan interval for such a beam-plot is typically one hydrophone diameter or less. If the beam is not symmetrical, then four beam-plots can be performed, each oriented at 45° to the previous one. The results from the four plots can then be combined to give the final result and the variation between the four plots can give an indication of the accuracy of the results. Fig. 7.10 gives a typical two-dimensional contour plot of the beam profile for a system where there is

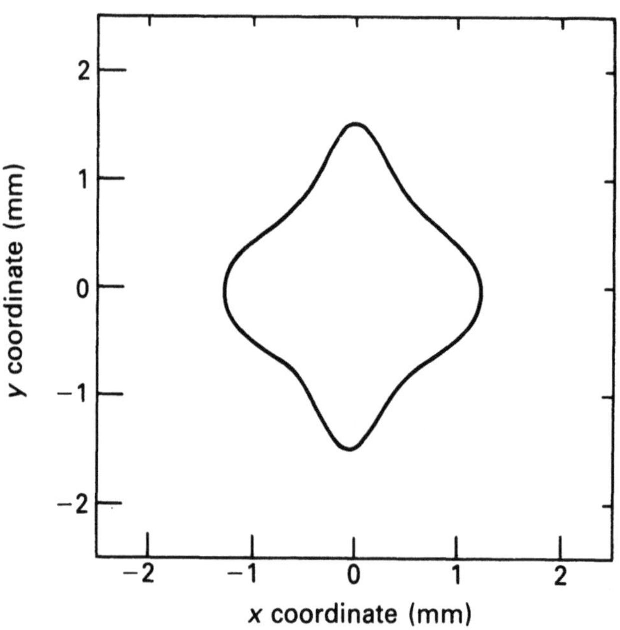

Figure 7.10. −6 dB contour plot of the pulse-pressure-squared integral in the focal plane. This diagram illustrates a typical case where the beam is not cylindrically symmetrical.

a different amount of focusing in the x and y directions, demonstrating the need to perform more than one beam-plot. In extreme cases of beam asymmetry, such as the field from a semi-circular transducer (as might be used in a split-crystal Doppler probe) a raster scan is probably the best option to choose.

Sometimes a scan is performed in order to calculate the total power in the beam (see 6.10). In this case, it is necessary to choose the scan limits. This could of course be decided in advance on the basis of the expected beam profile, but these expectations could be wrong. One can either scan out until the signal is less than a certain value, such as 10% of the peak, or the scan can be continued until the received signal is equal to the noise level of the measuring system. In either case, it is important to ensure that the scanning program does not confuse a local diffraction minimum within the beam with the true edge of the beam. To achieve this, the scan should be continued for several points after the apparent limit has been found, to ensure that the signal does not increase again.

7.4.2 Practical Details

During the scan, the hydrophone position and signal amplitude change considerably and this can lead to some practical difficulties. Firstly, the angle at which the incident wave meets the hydrophone changes with position and so, in principle, the hydrophone should be rotated for maximum signal at each step. This is usually too time-consuming and so one approach would be to ignore the effect and simply estimate the likely error that is introduced as a consequence. This estimate can be used in assessing the uncertainties, or it could be used to correct the results for the effect. An alternative method would be to estimate the angle of the incident beam (e.g. by assuming that it was propagating as a spherical wave) and rotate the hydrophone accordingly.

The changing amplitude of the waveform with position can make it difficult to obtain a trigger signal directly from the waveform. Probably the best method is to trigger from the waveform initially, but to change to a delayed trigger (initiated by the transducer drive pulse) after the signal has reduced in amplitude by a certain amount. The trigger time-delay may have to be varied, since the delay of the pulse will change with position. It is difficult to calculate this change in delay accurately, although it can be estimated from geometrical considerations. An alternative method would be to determine the change in delay between the pulse at the centre of the beam and that at the edge. The delay elsewhere in the beam could then be calculated by assuming that the change depends quadratically on the distance from the central axis.

7.5 Continuous-Wave Systems

These systems can be divided into continuous-wave (cw) Doppler and physiotherapy instruments and so will be dealt with accordingly.

7.5.1 CW Doppler Systems

As mentioned in 4.3.4, a problem common to any cw system is that of standing waves. These can be caused by reflections from the tank walls, hydrophone mount or the hydrophone itself. The first two types of reflection can be removed by placing absorbers in the appropriate positions. Possible types of absorber are domestic deep pile carpet (making sure to soak it thoroughly to remove air bubbles) or plates of paraffin wax. Further information on acoustical properties of some of these materials can be found in Chivers et al. [8].

It is not possible to eliminate the reflection from the hydrophone itself, but with membrane devices its effect can be reduced by rotating the hydrophone (Fig. 7.11a). If the angle is sufficient for the reflected wave to miss the edge of the transducer then the problem can be significantly reduced. The final results are then corrected knowing the directional response of the hydrophone and the rotation angle used. Sometimes it is not possible to eliminate these effects entirely, and in addition electrical interference at the hydrophone from the transducer can cause problems which resemble those of standing waves. In this case, the hydrophone should be moved a small distance towards or

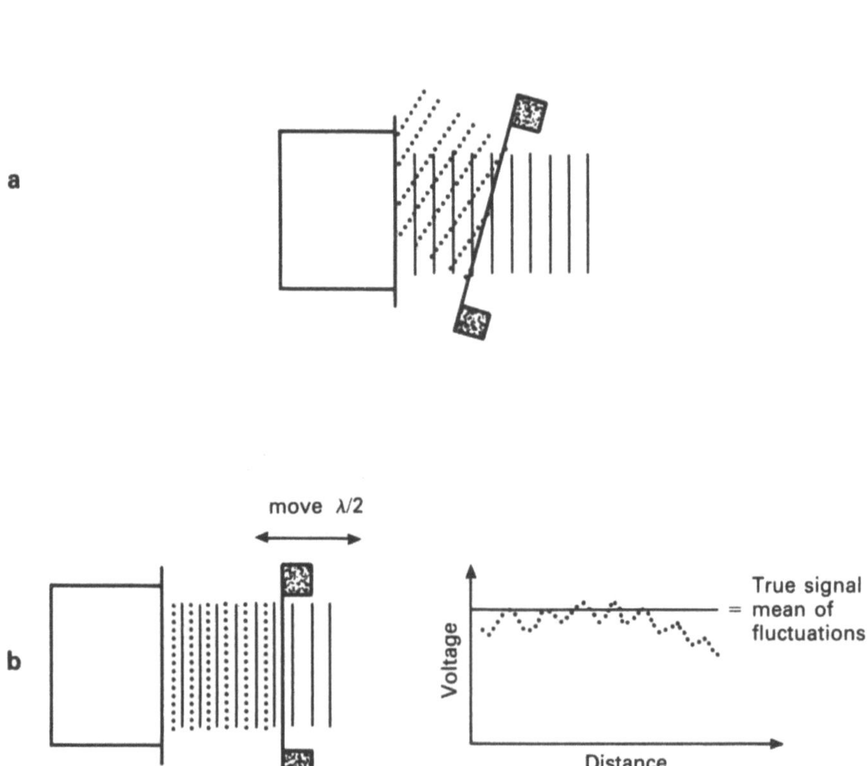

Figure 7.11. Overcoming the effects of standing waves. In **a** the hydrophone is rotated to avoid multiple reflections between it and the source. In **b**, the hydrophone is moved over a small axial distance to average out the fluctuations caused by standing waves or interference.

away from the transducer until a local maximum and then a local minimum of the signal are obtained. The true, free-field signal at that position is then given approximately by the average of the local maximum and minimum values (Fig. 7.11b).

With cw Doppler equipment it can be difficult to find the correct measurement position, because the beam may be diverging. If it can be shown by appropriate measurements that the maximum pressure occurs at the transducer face, then the US Food and Drug Administration (FDA) 510(k) Reporting Requirements [9] permit the spatial-average intensity to be calculated from the total ultrasonic power divided by the nominal transducer area. If the peak pressure occurs away from the transducer face, then the measurements have to be made at that position, using a hydrophone in the usual way.

Some Doppler systems (e.g. foetal heart detectors) use more than one source transducer. In this case, it is difficult to define a unique acoustic axis where the maximum pressure might be expected. The measurements may be made on the geometrical axis of the whole system. Under these circumstances, it is possible that larger acoustic pressures could be measured on the axis of one of the transducer elements.

7.5.2 Physiotherapy Equipment

The problem of standing waves has been dealt with in the previous section (7.5.1); the main additional feature of physiotherapy fields is their high amplitude and potential for producing cavitation. The first step to eliminate problems due to cavitation is to use clean, degassed water (see 4.3.1).

Measurements on physiotherapy equipment usually have to be performed very close to the transducer (5 mm from the face) and for this reason a hydrophone with good spatial resolution is required (see 7.2.2). Membrane hydrophones have relatively poor resolution at the frequencies of 0.7 to 3 MHz used for physiotherapy (see Eq. (7.1)) so a needle-probe hydrophone is to be preferred. Unfortunately, needle hydrophones can be very susceptible to cavitation damage, so the best approach is to reduce the drive amplitude so that cavitation does not occur. To assess the linearity of the instrument output a separate measurement can be made, possibly using a membrane hydrophone. To eliminate some of the measurement problems that are presented by the existing standards, work is under way to develop a new IEC standard for physiotherapy equipment. This will probably require measurements of total power using a radiation pressure balance and also beam profiles at greater distances from the transducer than are currently used.

Physiotherapy equipment can operate in the so-called pulsed mode as well as in continuous-wave mode. The pulses employed are typically very long (thousands of cycles), so that measurement problems due to standing waves may still be present. It is advisable to check that the peak pressure of the field in pulsed mode is the same as that for the equivalent continuous-wave mode. If this is the case, then it should be possible to relate the temporal-average intensities in the two modes simply by consideration of the appropriate duty factor.

7.6 Accuracy Assessment

An important aspect of the measurement of the acoustic output of medical ultrasonic equipment is assessing the experimental accuracy. If this is not done, then interpretation of the results becomes difficult. For instance, if two sets of results are compared (either from different laboratories or from the same laboratory at different times), one needs to know if any differences are significant and therefore a cause for concern, or whether they are within the experimental uncertainties. This decision could, for instance, affect whether or not an engineer is called in to examine the equipment for faults or even, perhaps, whether an item of new equipment is regarded as acceptable. These considerations are particularly relevant in the context of ultrasound output declaration and safety standards.

At a national standards laboratory such as NPL, considerable effort is put into this area, because a service is provided to customers, who rely heavily on the results. In a hospital situation, for instance, less effort can be justified, but nevertheless a simple assessment of the measurement uncertainties is required, for the reasons outlined above. The remainder of this section, therefore, considers some of the factors which give rise to inaccuracies in the measurements and which therefore must be considered. It also gives examples of the typical uncertainties that are currently quoted at NPL, which may be used as a guide. It must be stressed at this point, however, that each measurement is different and that the person making the measurements must be prepared to change the uncertainties that are quoted if this is appropriate. Examples of such situations would include a measurement where the output from the device under test changes during calibration, or where the beam size or operating frequency is outside the normal range.

There are two types of uncertainty that must be considered: the first is due to random fluctuations between measurements and is therefore called a random uncertainty (Fig. 7.12). The second type usually does not give rise to changes between repeated measurements because it is a function of the measuring system, and is called a systematic uncertainty (Fig. 7.13).

The random uncertainty can be assessed by repeating the measurements and noting any variations; these variations are then processed in a standard way to produce the random uncertainty at a given confidence level, which expresses the probability of the true result lying within the quoted uncertainty. A

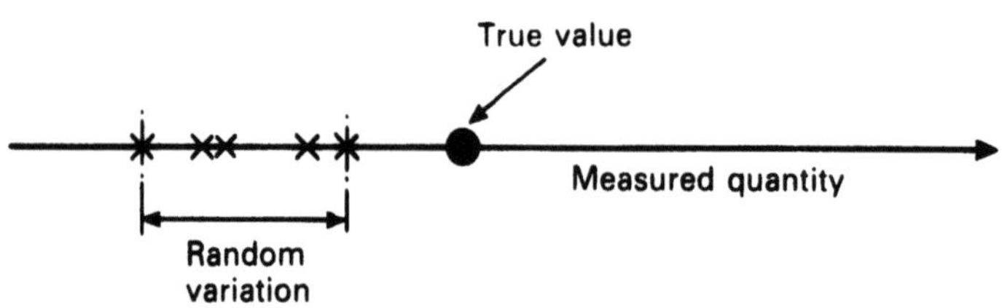

Figure 7.12. Illustration of the concept of random uncertainty, which is determined from the variation between repeated measurements of a given quantity. The measurement points are represented by crosses.

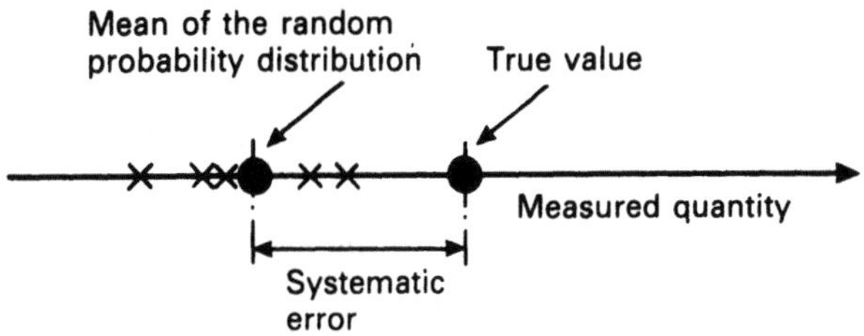

Figure 7.13. Illustration of the concept of systematic uncertainty. The mean of the measured data differs from the true value of a quantity by an amount which is known as the systematic error. This error cannot be determined directly, but its size has to be estimated. This estimate is known as the systematic uncertainty.

confidence level of 95% is generally used. If there are n different measurements, x_i, of the quantity x, with a mean denoted by $\bar{x}$, then the estimate of the random uncertainty at the 95% confidence level ($U_r(95\%)$) is:

$$U_r(95\%) = t_n \, [n(n-1)]^{-\frac{1}{2}} \left[\sum_{1}^{n} (x_i - \bar{x})^2 \right]^{\frac{1}{2}} \tag{7.10}$$

where t_n is the appropriate value of Student's "t" factor for the number of measurements n. t_n is equal to 12.7, 4.3, 3.18, 2.78 and 2.57 when n is 2, 3, 4, 5, or 6 respectively. For large values of n, t_n approaches an asymptotic value of 1.96. To obtain the uncertainty at different confidence levels, a different value for the "t" factor should be used; further details can be found in a leaflet available from the National Measurement Accreditation Service, National Physical Laboratory, Teddington, Middlesex, UK, entitled "The expression of uncertainty in electrical measurements", National Measurement Accreditation Service (NAMAS), Publication B3003.

By contrast with the random uncertainty, the systematic uncertainty cannot be estimated by repeating the measurements. Instead, it has to be evaluated by carefully considering the nature of the measuring system and the likely errors. One useful starting point in assessing this uncertainty is the manufacturer's specification for each item of equipment being used. In any real system, however, there are usually additional sources of uncertainty and these have to be identified and analysed separately by the user. When combining random uncertainties (Eq. (7.10)), it was implicitly assumed that the variation conformed to a normal probability distribution. For systematic uncertainties, a different (rectangular) probability distribution is assumed, so they are combined in a different manner. A conservative estimate of the overall systematic uncertainty $U_s(95\%)$ can be obtained as follows:

$$U_s(95\%) = 1.13 \left[\sum_{1}^{n} a_i^2 \right]^{\frac{1}{2}}, \tag{7.11}$$

where a_i are the estimates of the individual contributions to the systematic uncertainty. When one contribution is much larger than the others, the above formula can give a value that is greater than the sum of the individual a_i and so the true uncertainty is less than given above. The NAMAS publication referred to above gives the details of how to treat this case. To obtain the overall uncertainty $U(95\%)$, the random and systematic components should be combined in quadrature:

$$U(95\%) = [U_r(95\%)^2 + U_s(95\%)^2]^{\frac{1}{2}} \tag{7.12}$$

The first source of systematic uncertainty that should be considered is the calibration of the hydrophone and preamplifier. This should be quoted by the organisation performing the calibrations, since it cannot properly be assessed by anyone else. Secondly, one should consider the accuracy of the digitiser or oscilloscope. This should be quoted in the manufacturer's specifications, but the user must be careful to ensure that the accuracy figure employed applies at the typical signal levels and frequencies that are used. As an example, in assessing the accuracy of a digitiser in practice, one may have to account for the possibilities that the frequency may be close to the bandwidth limit of the device, that the signal may not extend over the whole screen and that different voltage ranges may have different accuracies.

The sensitivity of membrane hydrophones generally increases with temperature (by about 0.6% per °C for coplanar-shielded devices), which gives rise to an error if the hydrophone is used at a different temperature from that at which it was calibrated. Generally, this effect is small, but it should be accounted for if measurements are made above room temperature, e.g. at body temperature. In this case, the results could be corrected if desired, using published values for the temperature coefficient of membrane hydrophones [1]. Two further sources of uncertainty arise from limitations in the hydrophone

Table 7.1. Uncertainties (%) in the determination of various acoustical parameters using the NPL Beam Plotting Facility. These are typical figures for an automatic scanner operating at 3.5 MHz. All systematic uncertainties are semi-range values (i.e. a value of x% should be taken to read $\pm x$%)

Source	Uncertainty (%)										
	p_+	p_-	I_{sptp}	I_m	I_{sppa}	Pulse duration	I_{bpta}	I_{bata}	I_{spta}	Beam area	Beam power
Systematic											
Hydrophone/amplifier (95% confidence)	6.5	6.5	13	13	13	-	13	13	13	-	13
Temperature	0.6	0.6	1.2	1.2	1.2	-	1.2	1.2	1.2	-	1.2
Digitiser	2	2	4	4	4	-	4	4	4	-	4
Spatial averaging	3	1	5	6	4	1	4	3	2	4	1
Nonlinear distortion	8	0	16	9	4	1	4	3	2	4	1
Overlapping scan lines	-	-	-	-	-	-	-	-	15	-	-
Total systematic (95% confidence)	12	7	24	19	16	2	16	15	22	7	14
Random											
Random uncertainty (95% confidence)	2	2	4	3	3	1	4	7	4	5	4
Overall uncertainty (95% confidence)	13	8	25	20	17	3	17	17	23	9	15

Table 7.2. Typical values for the uncertainties in the determination of various acoustical parameters using the NPL Ultrasound Beam Calibrator for an automatic scanner operating at 3.5 MHz. All systematic uncertainties are semi-range values (i.e. a value of x% should be taken to read $\pm x$%)

Source	Uncertainty (%)							
	p_+	p_-	Beam width	Pulse duration	I_{sppa}	I_{spta}	I_{bata}	Beam power
Systematic								
Calibration	8.5	8.5	–	–	17	17	17	17
Bandwidth	14	–	13	4	19	9	19	4
Digitisation	2	2	4	4	4	4	4	4
Gain	7	7	7	–	14	14	14	14
Spatial averaging	5	2	10	2	8	8	8	2
Pulse repetition rate	–	–	–	–	0.01	0.01	0.01	0.01
Hydrophone tilt	0.4	0.4	0.4	–	0.8	0.8	0.8	0.8
Cylindrical symmetry	–	–	–	–	–	–	9.5	45
Total systematic (95%)	21	13	21	7	35	29	36	57
Random								
Random (95%)	4	2	1	1	2	2	2	3
Overall (95%)	22	13	22	8	36	30	37	58

response and have already been discussed in 7.2; they are the spatial averaging of the field by the hydrophone and the effect of the limited bandwidth of the hydrophone. Indications of the uncertainties that these effects introduce have already been given; further information is given by Smith [4,5].

Tables 7.1 and 7.2 give the typical uncertainties quoted at NPL for the Beam Plotting Facility and the Ultrasound Beam Calibrator respectively. The uncertainties quoted in Table 7.2 would, of course, normally apply to all users of the Ultrasound Beam Calibrator. It should be noted that the tables quote a different uncertainty for each acoustical parameter; this is important because the values can be very different for each parameter. In Table 7.1 there is an additional uncertainty that has not been discussed here, namely that due to "overlapping scan lines". This is outside of the scope of the present discussion but is described in 9.3.6. Table 7.2 includes an uncertainty for the lack of cylindrical symmetry in the beam; this arises because the Ultrasound Beam Calibrator does not perform a full raster scan to measure the spatial-average acoustical quantities and so has to assume a certain degree of cylindrical symmetry in the beam. For a cylindrically symmetrical beam the uncertainty for beam power would be considerably smaller than that given in Table 7.2.

7.7 Recommended Additional Reading Material

Duck FA. The measurement of exposure to ultrasound and its application to estimates of ultrasound "dose". Phys Med Biol 1987; 32: 303–325

Harris GR. A discussion of procedures for ultrasonic intensity and power calculations from miniature hydrophone measurements. Ultrasound Med Biol 1985; 11: 803–817

Harris GR. Hydrophone measurements in diagnostic ultrasound fields. IEEE Trans Ultrason Ferroelec Freq Contr 1988; UFFC-35: 87–101

Preston RC. Measurement and characterisation of the acoustic output of medical ultrasonic equipment (Parts 1 and 2). Med Biol Eng and Comput 1986; (Part 1) 24: 113–120; (Part 2) 24: 225–234

Szabo TL, Melton HE, Hempstead PS. Ultrasonic output measurements of multiple-mode diagnostic ultrasound systems. IEEE Trans Ultrason Ferroelec Freq Contr 1988; UFFC-35: 220–231

7.8 References

1. Preston RC, Bacon DR, Smith RA. Calibration of medical ultrasonic equipment – procedures and accuracy assessment. IEEE Trans Ultrason Ferroelec Freq Contr 1988; UFFC-35: 110–121
2. Martin K. Measurement of acoustic output parameters from medical ultrasound devices with an RF power meter system. IEEE Trans Ultrason Ferroelec Freq Contr 1988; UFFC-35: 140–145
3. Preston RC. The NPL ultrasound beam calibrator. IEEE Trans Ultrason Ferroelec Freq Contr 1988; UFFC-35: 122–139
4. Smith RA. The importance of the frequency response of a hydrophone when characterising medical ultrasonic fields. Proc Inst Acoust 1986; 8: 118–128
5. Smith RA. Are hydrophones of diameter 0.5 mm small enough to characterise diagnostic ultrasound equipment? Physics Med Biol 1989; 34: 1593–1607
6. IEC. Measurement and characterisation of ultrasonic fields using hydrophones in the frequency range 0.5 MHz to 15 MHz. To be published as an IEC Standard, Document 87(CO)6, International Electrotechnical Commission, Geneva, Switzerland, 1988
7. Beissner K. Maximum hydrophone size in ultrasonic field measurements. Acustica 1985; 59: 61–66
8. Chivers RC, Smith AD, Filmore PR. Broadband measurements on ultrasonic tank lining materials. Ultrasonics 1981; 19: 125–133
9. FDA. Guide for measuring and reporting acoustic output of diagnostic ultrasound medical devices. Document 510(k), US Dept of Health and Human Services, Food and Drug Administration, HFZ-132, Rockville, MD 20857, USA, 1985

8

Real-Time Scanning Systems Part 1: Modes of Operation

A. SHAW

8.1 Introduction

Chapters 2 to 5 dealt generally with the basics of the propagation of ultrasound and with the different ways of measuring ultrasonic fields; chapters 6 and 7 concentrated on using that knowledge to characterise pulses of ultrasound. This and the next chapter consider how to characterise not a single pulse of ultrasound but the acoustic output of a real-time scanner.

This opens up a whole new set of difficulties, not the least of which is the need to set up the scanner and position the measuring hydrophone in such a way that we are really measuring the maximum value of I_{spta}, p_+ or whatever. To do that you obviously need to understand how the scanner works and what control you have over the acoustic output; this is the topic that is dealt with in this chapter.

Four main areas will be considered:

1. Principles of operation: how does a scanner build up a picture of the patient?
2. Pulse characteristics: what types of acoustic pulse are used for different applications? Which parameters are likely to be the most important in each case, and what are typical values?
3. Pulse patterns: what is the sequence of pulses fired in the different operating modes?
4. Scanner settings: which scanner controls are likely to affect the output and how?

Modern scanners are very complicated machines and they are all different, so it is not possible to look at all the variations; nevertheless, this chapter aims to give some insight into their operation.

8.2 Principles of Operation

When ultrasound travels from one type of tissue to another – say from fat into muscle, or from healthy liver into cancerous liver – some of the ultrasound is reflected from the interface. In addition, blood and tissue contain inhomogeneities which scatter part of the ultrasound back in the direction it has come from. Scanners make use of these two phenomena to construct a picture of the tissue under examination.

The transducer emits a pulse of ultrasound in a well-defined direction and then waits for the reflected and scattered sound to come back. The transducer then acts as a receiver and converts the acoustic pressure to a voltage. The elapsed time between transmission and reception of the pulse is proportional to the distance to the reflecting object; the strength of the reflected pulse is governed by the acoustic impedance mismatch at the object boundary. Typically, a boundary between two different types of tissue will give relatively strong reflections, but variations within a single tissue type give rise to much weaker signals. An imaging scanner makes use of the received amplitude to build up a picture; Doppler scanners use the change in frequency of the wave as it returns from moving structures or circulating blood to determine their velocity.

There are many different designs of transducer, such as linear array, mechanical sector scanner (or "wobbler") and phased array, some of which are illustrated in Fig. 8.1. In each case the principle is the same but different methods are used to steer and to focus the ultrasound beam.

In the plane of the image (or "scan plane") the beam-width is determined by the shape of the radiating element (static focusing), or by the number of elements used to produce each pulse and the electrical phase delays introduced between the elements (dynamic focusing). The beam narrows as it moves from the transducer to the focus, and spreads again beyond the focus.

Perpendicular to the scan plane the beam-width is not actively controlled but is usually determined by the geometry of the transducer, including the effect of any acoustic lens that might be present, and by diffraction. In consequence, the narrowest part of the beam may not occur at the same distance in each of the two directions. An exception is annular arrays which have concentric ring-shaped elements and produce cylindrically focused beams.

8.2.1 Imaging

In the simplest form of one-dimensional imaging systems each pulse of ultrasound is fired in the same direction. When the display is of reflected amplitude versus depth (like the trace on an oscilloscope) it is called A-mode; this was the earliest form of imaging. More common now is M-mode, where depth is displayed on the vertical axis and the reflected amplitude is used to modulate the brightness of the image. The images from successive acoustic pulses are stacked horizontally, giving a display of position against time and allowing the ultrasonographer to see the motion (hence M-mode) of a particular feature. Acoustically, both these modes are identical – they will both be referred to as M-mode – and making measurements on such a system is exactly

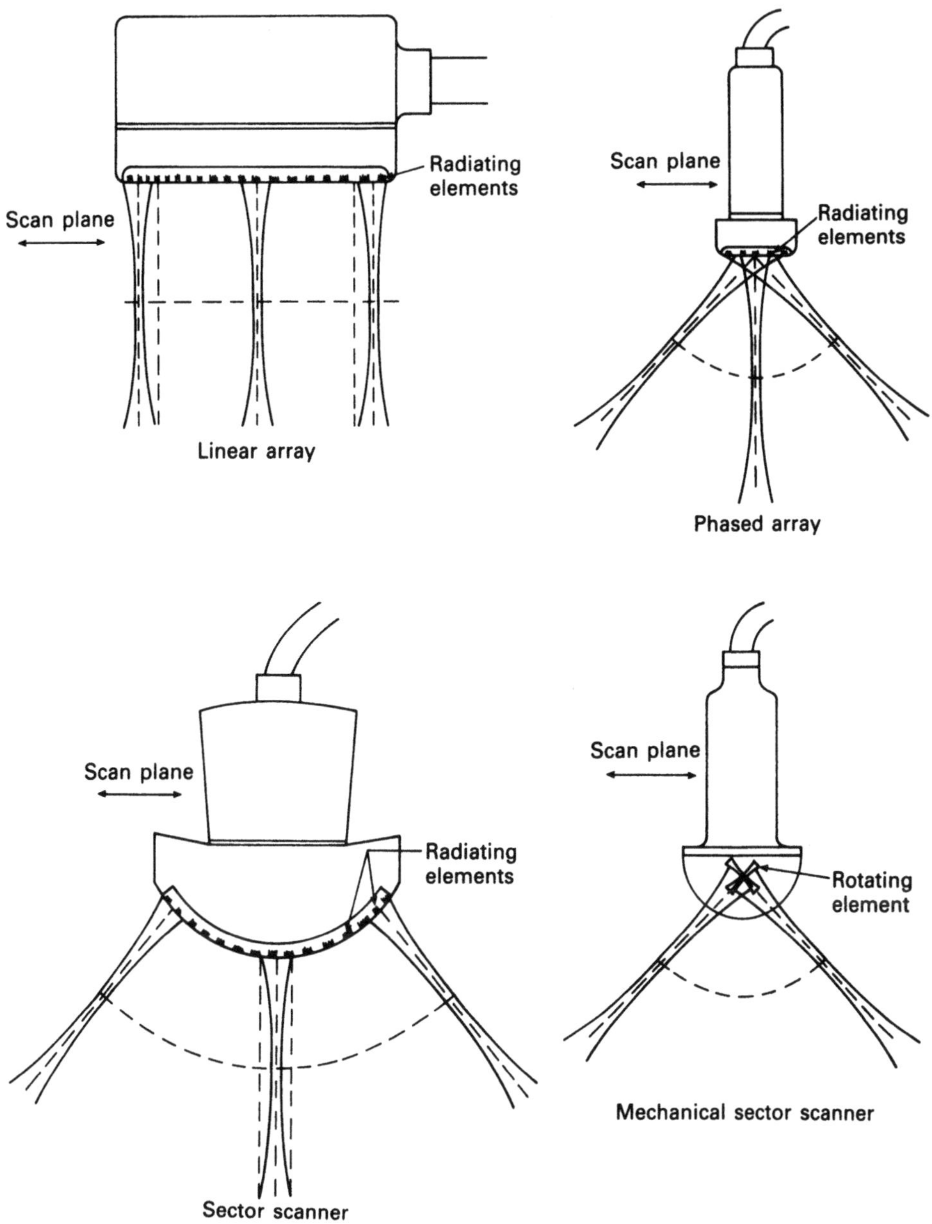

Figure 8.1. Different transducer geometries.

like characterising the beam from a single-element transducer as described in Chapters 6 and 7.

Building up two-dimensional information was originally achieved using the static B-scanner in which the transducer was moved systematically across a patient and a position registration system used to control the position of the modulated line image on a display screen, thereby building up what is termed a B-mode display. The systems shown in Fig. 8.1 build up a two-dimensional image of a slice of tissue by firing a number of ultrasonic pulses, each in a different direction within the scan plane, and waiting for the echoes from each direction in turn. The pattern repeats after a set number of pulses; each set of pulses is called a "frame" and the number of frames generated each second is the "frame repetition rate".

Making measurements on a system like this is different to measuring an M-mode scanner. Firstly, because each beam is fired in a different direction or from a different starting position, only one beam in each frame will be aimed at the measuring hydrophone. In other words, it is not good enough to look at the waveform from every acoustic firing, you need to pick out one pulse in each frame and lock on to it. This is difficult enough when looking for the spatial-peak parameters, but even more so when trying to build up a profile of the beam. As the hydrophone is moved off axis, the signals from neighbouring scan lines may become bigger than the signal of interest (see Fig. 8.2), making triggering and synchronisation difficult. These problems are dealt with in Chapter 9. However, once you have locked-on to the right scan line it should be possible to characterise that beam in the same way as described in Chapters 6 and 7 and to calculate values of p_+, p_-, beam-width, pulse-average pressures and intensities, and so on.

The second major new difficulty now comes in working out the temporal-average values for pressure and intensity. To do this it is necessary to measure the contribution of other scan lines, or "overlap" (see Fig. 8.3). This means that as well as triggering on one particular scan line you must be able to trigger on each neighbouring line as well. In principle the contribution from every scan line in the frame should be included, although it may be possible to simplify the process by assuming that many have a negligible influence, as explained in Chapter 9.

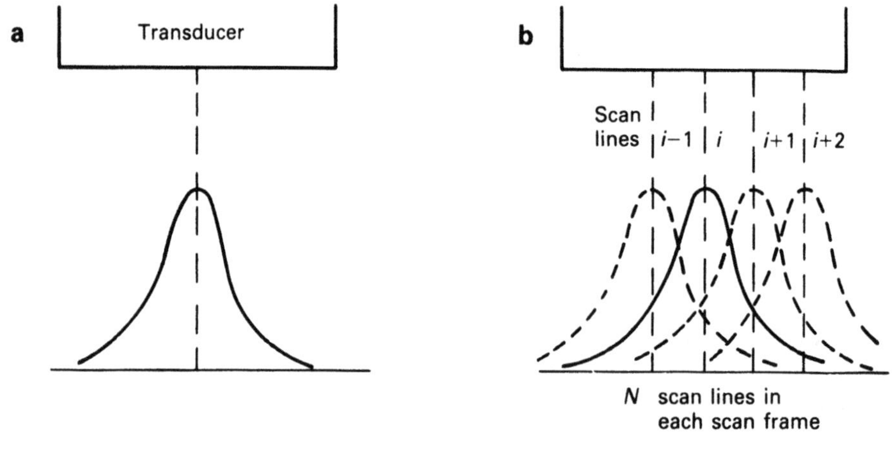

Figure 8.2. **a** M-mode: every pulse is exactly equivalent. **b** B-mode: not all pulses are equivalent; you must lock-on to the ith pulse in each frame.

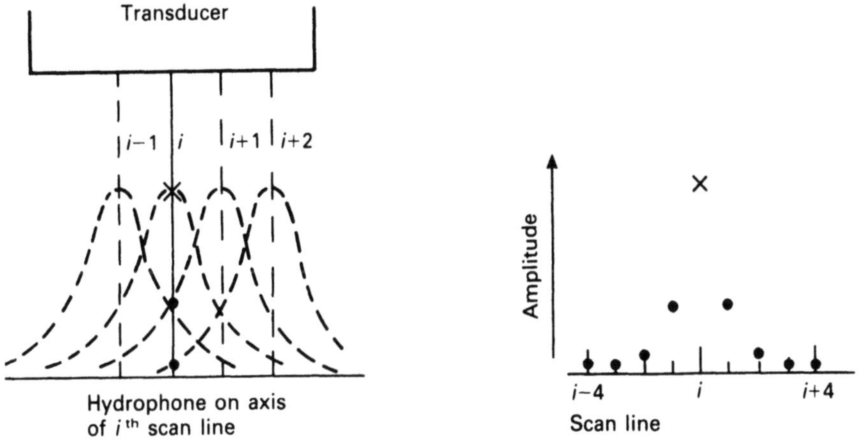

Figure 8.3. Investigating the overlap in a linear array.

8.2.2 Doppler

The main use of Doppler ultrasound is to measure the speed and quantity of bloodflow. The technique requires measurement of the frequency of the scattered/reflected wave and comparison with the frequency of the transmitted wave. The change in frequency is proportional to the component of the velocity of the blood towards or away from the transducer.

Doppler beams are focused in the same way as imaging beams, i.e. by both static and dynamic means. A single steerable Doppler beam is normally used (D-mode). The exceptions to this are continuous-wave (cw) Doppler, which is generated by a separate fixed beam transducer, and "colour-flow Doppler", which interlaces Doppler and imaging lines throughout the scan plane. Continuous-wave beams are not usually focused dynamically, although in principle there is no reason why they should not be.

8.3 Pulse Characteristics

After sorting out the triggering and locking-on to the right scan line what can you expect to see on your oscilloscope? This section will look at typical ultrasound beams that might be used for imaging, for pulsed Doppler, and for continuous-wave Doppler; which exposure parameters are likely to be the most important for safety, and what values might these parameters have?

8.3.1 Imaging

Fig. 8.4 shows a fairly typical waveform close to the position of maximum p_+. It is asymmetric and is greatly distorted due to the effects of nonlinear

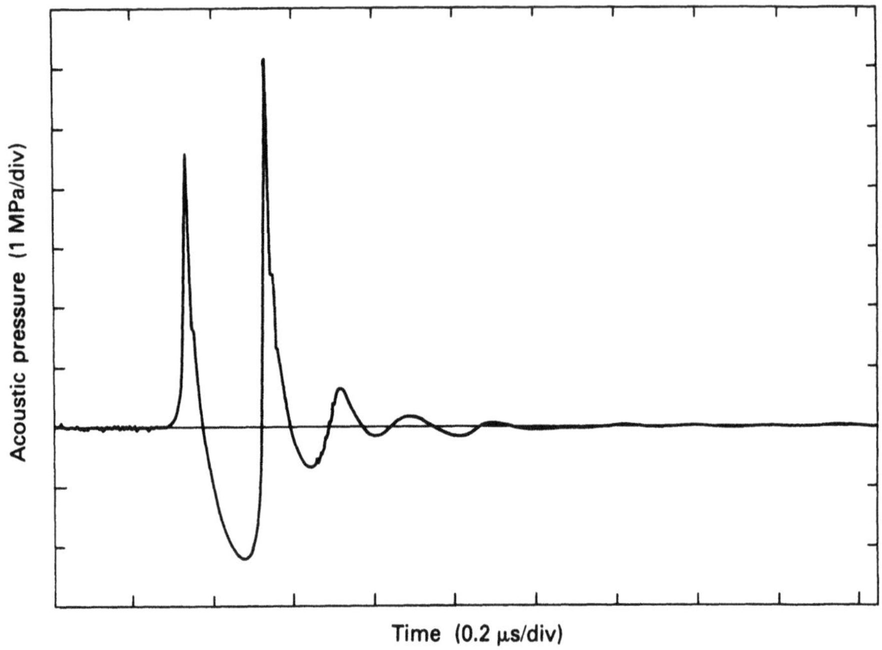

Figure 8.4. A typical distorted M-mode or B-mode waveform.

propagation, as mentioned in 7.2.1; the degree of distortion will vary with the pressure amplitude and the position of the focus. Typically, the acoustic working frequency is between 2.25 MHz and 7.5 MHz and the pulse contains one or two cycles. The beam-width is generally of the order of 1.5 mm but may be as small as 0.4 mm for some specialised skin-scanners.

Peak-positive pressures for imaging beams may be as large as 9 MPa when measured in water but are usually between 2 and 5 MPa; peak-negative pressures are usually between 1 and 3 MPa. Because of the high pressures there may be a risk of cavitation and consequent cellular damage. Pulse-average intensities are also very high, but note that this does not necessarily mean that the temporal-average values will be excessive. For a scanner operating in B-mode, the beam is not stationary but is swept, "smearing out" the energy over a large area. The highest values of I_{spta} will occur in M-mode when every acoustic pulse is fired in the same direction; in this case the energy absorbed in the tissue could produce an undesirable temperature rise of more than 1°C. Typical values for parameters measured near the focus are given in Table 8.1; although most scanners will give results in the ranges indicated, it is not uncommon to discover higher or lower values.

8.3.2 Pulsed Doppler

For Doppler systems the frequency of the transmitted wave must be well defined; the pulse usually contains at least three, and often ten or more cycles. Fig. 8.5 shows a typical Doppler waveform near the focus; peak pressures are

Table 8.1. Typical parameter values for diagnostic equipment

	M-mode	B-mode	Pulsed Doppler	CW Doppler PV	CW Doppler foetal
p_+ (MPa)	2–5	2–5	2–5	0.02–0.1	~0.02
p_- (MPa)	1–3	1–3	1–3	0.02–0.1	~0.02
−6 dB beam-width (mm)	1–3	1–3	1–3	~4	~20
Pulse-duration (μs)	0.2–1.0	0.2–1.0	0.3–10	cw	cw
I_{spta} (mW/cm^2)	50–200	1–30	300–5000	20–600	~20
I_{sppa} (W/cm^2)	50–500	50–500	50–500	–	–
Total power (mW)	1–100	1–100	20–250	5–50	~15

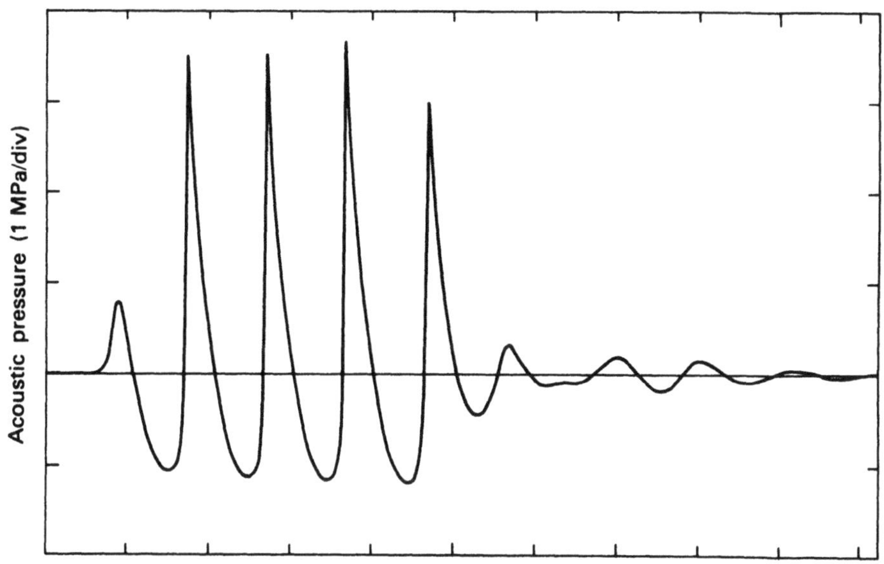

Figure 8.5. A typical distorted pulsed Doppler waveform near the focus.

similar to those for imaging pulses (p_+ from 2 to 5 MPa, p_- from 1 to 3 MPa) so cavitation is again a possible adverse effect. The combination of high pressure, longer pulse duration and high pulse repetition rate means that I_{spta} values are even higher for Doppler than for M-mode (typically 300 to 5000 mW/cm^2) so there is undoubtedly a risk of significant heating of tissue. Typical values for the parameters are given in Table 8.1.

8.3.3 Duplex

Duplex is a combination of imaging and pulsed Doppler. This means that there are two different types of acoustic pulse, as described in 8.3.1 and 8.3.2, which have to be characterised separately. I_{spta} is calculated by summing the contributions from both sets of pulses.

8.3.4 CW Doppler

The pressures for cw Doppler are very much lower than for imaging or pulsed Doppler (typically p_+ and p_- are less than 0.1 MPa), consequently there is little risk of cavitation effects. However, because the transducer is active all the time, and because the beam is stationary, the temporal-average intensities can be high, maybe 20 to 600 mW/cm^2.

Typical parameter values are again given in Table 8.1 and the waveform illustrated in Fig. 8.6. There are two distinct classes of cw Doppler: peripheral vascular (PV) or cardiovascular, and foetal monitor. Foetal monitors generally lie at the lower end of the scale, especially for I_{spta} and total power. Also, because of the need to cover a large tissue volume, foetal monitors usually have large and divergent beams; this means that the maximum values of pressure and intensity lie close to the transducer.

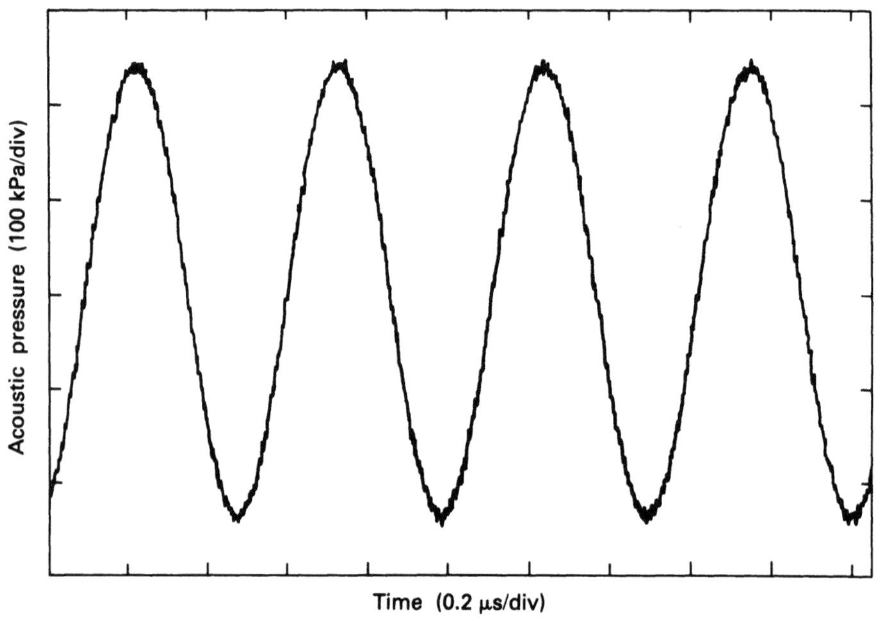

Figure 8.6. A sample of a typical cw Doppler waveform.

8.4 Pulse Patterns

Assuming that it is possible to lock-on to one scan line and to position the hydrophone on the axis of that line, what would you now expect to see if, without moving the hydrophone, you were to look in turn at the contributions from each of the other scan lines?

The best way to show this is with the type of diagram used in Fig. 8.3. The vertical scale is the amplitude of the acoustic signal (p_+ in this case) – be it a short pulse or toneburst. The horizontal axis depicts the arrival of consecutive scan lines. (It is possible to get this sort of display by using a slow time-base on your oscilloscope and triggering on the start-of-frame pulse. In this case the horizontal axis will be time.) The number assigned to each scan line has no particular significance here except that 0 is the reference, −3, −2, −1 are the lines before scan line 0, and 1, 2, 3 are the lines transmitted afterwards.

This section will look at cases of increasing complexity and explain the patterns that occur. Naturally, it will not be possible to cover all the feasible arrangements, but it will demonstrate how to gain an understanding of the operating mode of a complicated piece of equipment. This understanding is one of the most important aspects of making measurements on real-time systems. Without it the measurements are much more difficult to interpret.

Scanners can operate in a range of discrete (single) modes or in combined (also called mixed or multiple) modes, and some of the more common modes will be examined in the following sections. A knowledge of pulse pattern is also vital in order to determine the total acoustic power output.

8.4.1 Discrete Modes

M-Mode and D-Mode.

Fig. 8.7 illustrates the simplest case, in which every pulse is the same and is directed at the hydrophone. This is what you would see if you were looking at a single element transducer, a scanner operating in M-mode only, or a scanner operating in pulsed Doppler (D-mode) only. To distinguish D-mode from M-mode you need to look at the pulse waveform (see 8.3).

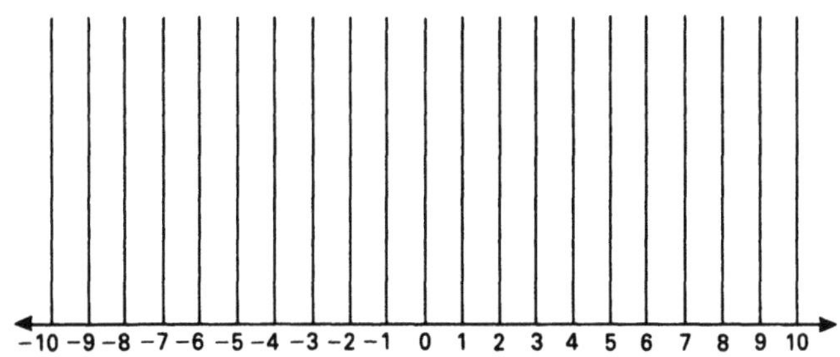

Figure 8.7. A typical M-mode or D-mode pattern.

B-Mode.

Scanning systems generate a sequence of ultrasonic pulses, with each pulse being fired in a different direction. When the beam is directed away from the hydrophone there will be no signal detected. As the beam gets closer the signal will increase until it reaches a maximum at scan line 0 and then fall off again, as is shown in Fig. 8.8.

If the beam is very tightly focused, and the hydrophone is located at the focus, it is possible that only the central line and the two on either side (+1 and −1) will give a signal at the hydrophone (Fig. 8.9).

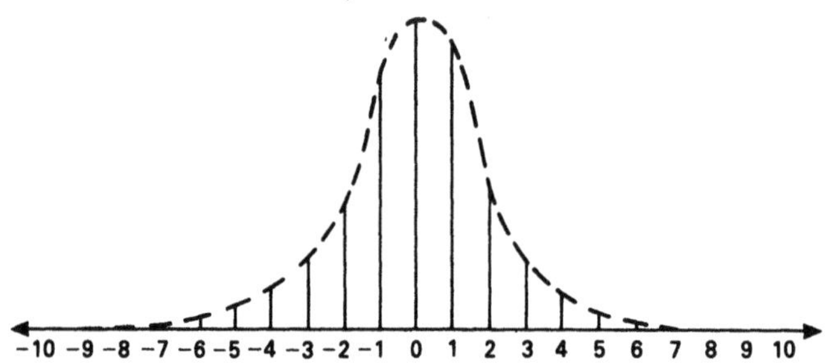

Figure 8.8. A typical B-mode pattern.

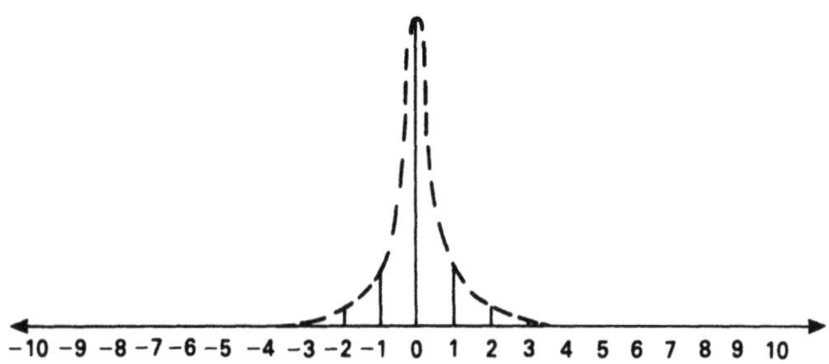

Figure 8.9. A sharply focused B-mode pattern.

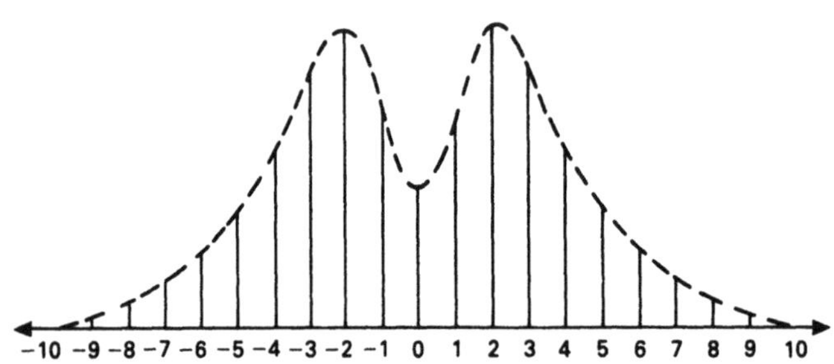

Figure 8.10. A typical near-field B-mode pattern.

If the hydrophone is in the near-field of the transducer the point of highest pressure may not correspond to the geometric centre of the beam, and a pattern similar to that shown in Fig. 8.10 may be observed.

8.4.2 Combined Modes

B+M-Mode.

In B+M-mode the scanner generates B-mode imaging lines to build up a picture of the tissue structures; in addition, each frame includes a number of

M-mode lines to display movement. The imaging lines are needed to provide the ultrasonographer with positional information and relate the M-mode line to the surrounding tissue structures. It is usually possible to move the M-mode line within the image to pick out different structures.

The way the two modes are interlaced and the effect of introducing M-mode upon the number of scan lines in each frame will vary from scanner to scanner. One method used is to introduce an additional M-mode line after every fifth pulse (say) and build up the imaging frame using the other pulses.

Figs. 8.11 and 8.12 illustrate B+M-mode. The former shows the case where the M-mode lines are directed at the hydrophone so, in addition to the normal B-mode pattern, every sixth pulse gives a large signal. In Fig. 8.12, however, the large M-mode signals are replaced by "holes", or zero signals, because the M-mode line has been moved a long way from the hydrophone. Obviously there will also be intermediate cases.

B+D-Mode.

B+D-mode is generally known as Duplex. Introducing a pulsed Doppler line into an imaging frame could also give the patterns in Figs. 8.11 and 8.12, but with the M-mode lines replaced by Dopper lines. The amplitude of the Doppler and imaging pulses will probably be different. It is also possible to produce a

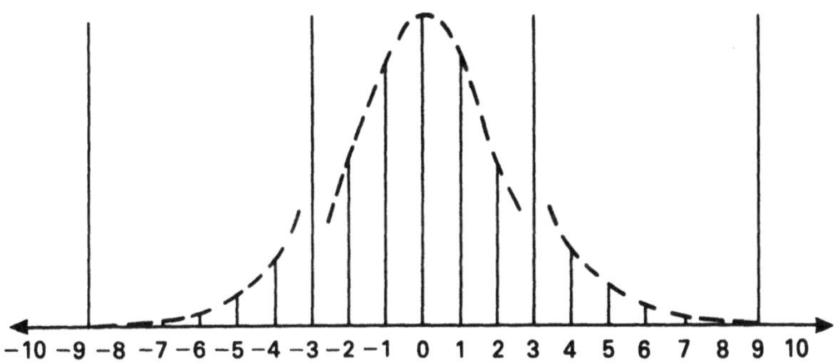

Figure 8.11. A typical B+M-mode pattern on the M-mode axis.

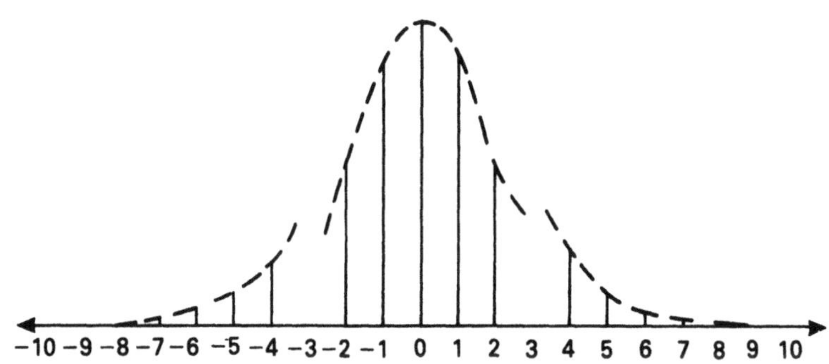

Figure 8.12. A typical B+M-mode pattern off the M-mode axis.

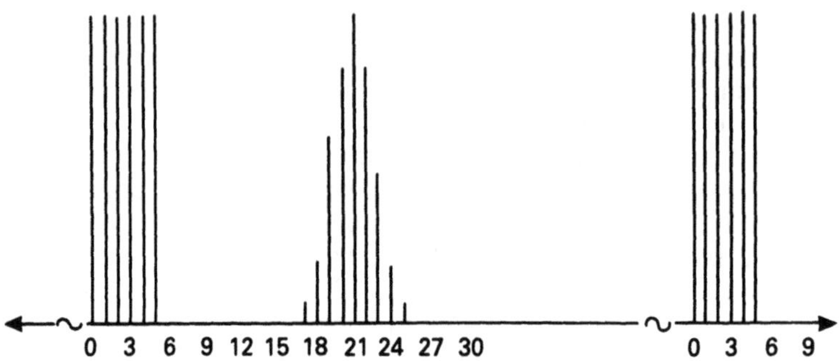

Figure 8.13. A possible B+D-mode pattern.

wide range of other patterns by interlacing the lines in different ways, one example is given in Fig. 8.13. Here, six Doppler lines are followed by a set of imaging lines.

B+D+M-Mode.
Definitely one for the expert! Imagine a mixture of B+M and B+D modes; almost any combination is possible. The way to cope with this is to look for the profile of the imaging lines and work from there. In practice, it should be possible to operate the system in simpler modes to build up information for this complex mode.

8.4.3 Multiple Focal Zones

It is often possible for a scanner to fire two or more sets of imaging lines in a frame, with each set of lines focused at a different distance. Each set is referred to as one "focal zone". For example, focusing all the odd lines at one distance, and all the even lines at another will generate two focal zones. This can be applied to all the operating modes.

M-Mode and D-Mode.
In either M-mode or D-mode, the peak pressures at the hydrophone due to each focal zone will generally be different, giving the pattern in Fig. 8.14.

B-Mode.
Multiple focal zones in B-mode can give very confusing patterns as shown in Fig. 8.15, especially if there are more than two zones. The trick is to try and pick out the profile from each zone if you suspect this is what you are seeing.

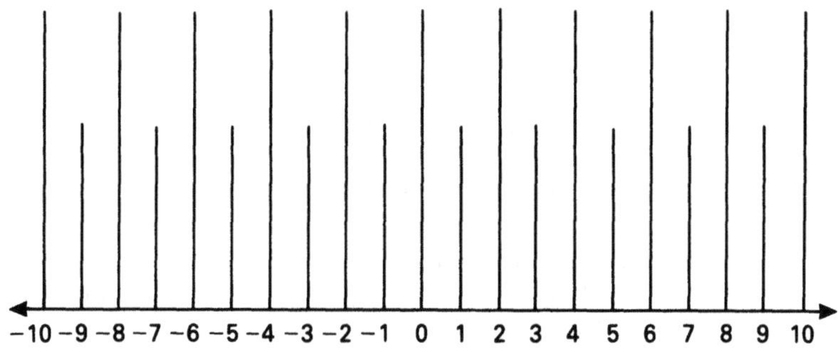

Figure 8.14. A typical M-mode or D-mode pattern for two focal zones.

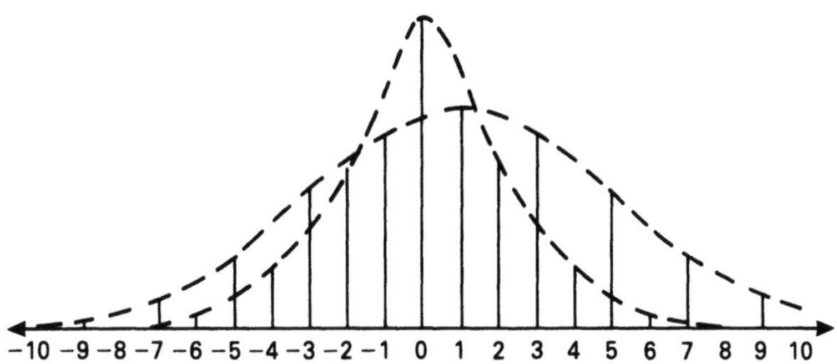

Figure 8.15. A typical B-mode pattern for two focal zones.

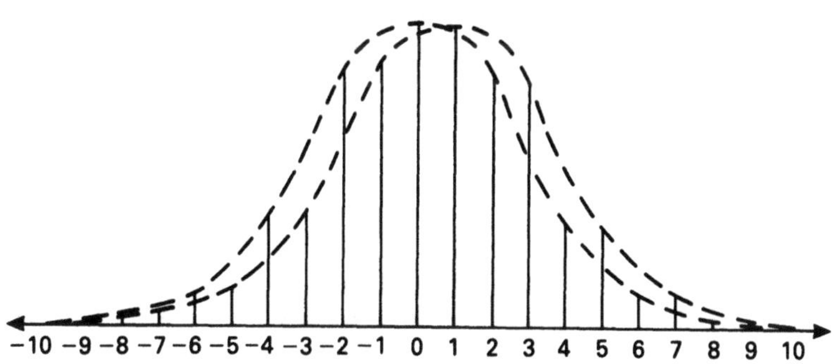

Figure 8.16. A typical repeated B-mode pattern.

8.4.4 Repeated Lines

Repeated lines should not be confused with multiple focal zones. Repeated lines are two or more sets of identical lines as shown in Fig. 8.16. This is achieved by simply duplicating each acoustic pulse. Again, duplicate lines may appear in any of the scanner modes.

8.5 Scanner Settings

Broadly speaking there are usually four characteristics of scanner acoustic output that can be varied from the front panel: the pressure amplitude of the ultrasound, the pulse duration, the pulse repetition rate (*prr*) and the frame repetition rate. The last two are obviously related and will be dealt with together. Although these parameters are referred to here in acoustical terms, they are controlled by settings which do not necessarily correspond to these terms, as seen in 8.5.1 and 8.5.2.

On some machines it is possible to vary one or more of these acoustical parameters independently and in a logical manner. However, controls are not always independent, especially on the more advanced scanners, and altering one control may have unexpected effects on another aspect of performance. In order to determine the maximum possible exposure when characterising a scanner you must investigate the effects of all the front panel controls, and be able to evaluate rapidly any changes. Instruction manuals are not generally very useful in explaining the effects on the acoustic output.

8.5.1 Imaging

Pressure Amplitude

Generally set by an overall power control. It is very important that this is set to maximum when trying to find the maximum exposures. Also, different focal zones will probably generate different acoustic powers; they will certainly affect the amplitude of the pulse at the measurement point.

Note: Power is not the same as "brightness" and not the same as "gain", both of which usually affect only signal processing or the displayed image.

Pulse Duration

Not normally under operator control.

Pulse Repetition Rate

Pulse repetition rate (*prr*) is often automatically controlled by the selection of the depth on the displayed image. If full depth is required then a low *prr* is necessary to avoid multiple echoes. When imaging near the transducer face a higher *prr* may automatically be selected. In M-mode there is often a control specifically for *prr*.

Frame Repetition Rate

This can vary with the size of the displayed image, or "scale" (also called "magnification"), and the minimum displayed depth, or "scroll" position. The number of focal zones being fired may also have an effect. In B-mode, sector scanners and phased arrays may have a variable sector angle which is likely to affect the frame rate.

8.5.2 Pulsed Doppler

Pressure Amplitude
There may be a Doppler power control (again note the distinction between power, gain and brightness). Amplitude may also be affected by altering the length of the Doppler pulse (see next paragraph); the amplitude is sometimes automatically reduced as the pulse length increases, in an attempt to keep total power constant. If it is possible to select different focusing regimes then that will also affect the pulse amplitude.

Pulse Duration
This is changed by altering the "gate width" or "sample volume"; if the peak pressure is not affected then the greatest "gate width" will give maximum I_{spta}. However, amplitude and pulse duration are often related and the pulse-pressure-squared integral should be evaluated for each setting.

Pulse Repetition Rate
Prr may be affected by the imaging frame rate but there may also be a separate Doppler control (often the "range gated depth" which controls the depth of the sample volume within the tissue). Generally the highest *prr* will give the maximum temporal-average intensities; but again there may be a link between *prr* and amplitude.

8.5.3 Duplex

In any mode, including colour-flow, which combines imaging and pulsed Doppler beams it may be possible to alter each type separately (see 8.5.1 and 8.5.2). Often, however, the two will not be independent and you will need to investigate the various controls to discover how they affect output.

8.5.4 CW Doppler

The power should be set to maximum; there will probably be no other controls.

8.6 Discussion

This chapter has looked at the principles of how a scanner works, what type of acoustic output you can expect, and some of the extra difficulties that you will come up against when trying to characterise a real-time system. It has not examined ways of handling these difficulties which will be covered in the next chapter.

The purpose here has been to make it clear that modern scanners can be extremely complicated, but that given a knowledge of the way the system works, it is quite feasible to undertake measurements. Many of the problems addressed here have been solved by the development of the NPL Ultrasound Beam Calibrator which is a commercially-available measurement system designed to determine the acoustic output of these complex systems. Its use has, for the first time, made acoustic data from a large number of ultrasonic scanners widely available.

8.7 Recommended Further Reading Material

Fish P. Physics and instrumentation of diagnostic medical ultrasound. John Wiley and Sons, Chichester, UK, 1990

McDicken WN. Diagnostic ultrasonics: principles and use of instruments. John Wiley and Sons, New York, 1981

9

Real-Time Scanning Systems Part 2: Measurements

R.C. Preston

9.1 Introduction

This chapter deals with the determination of the acoustic output of an ultrasonic scanner – in other words a system which is generating a non-static beam of ultrasound. Such systems could be mechanical sector, linear- or phased-array scanners. It is assumed here that the only types of measurement being made are those using hydrophones.

9.2 Measurement Systems

9.2.1 Test-Tank Arrangements

Depending on the arrangement of the test-tank, precautions will need to be taken to ensure water does not penetrate the housing of the transducer. Manufacturers advice is normally provided on the degree of immersion allowed in water. Apart from intra-operative and intra-cavity probes, it is not possible to immerse most probes in water to a depth more than 1–2 cm from the front face. Special precautions will need to be taken if more extensive immersion is necessary.

There are various basic arrangements for making measurements of the output using a hydrophone and scanning tank. These were illustrated in Chapter 7 and are shown again in Fig. 9.1. In the first the transducer is arranged so that only its face is immersed in water and the ultrasound is directed vertically downwards. This is the simplest method and is widely used, as in the NPL Ultrasound Beam Calibrator [1]. The only precaution necessary here is to ensure the immersion depth is within the manufacturer's recommendations.

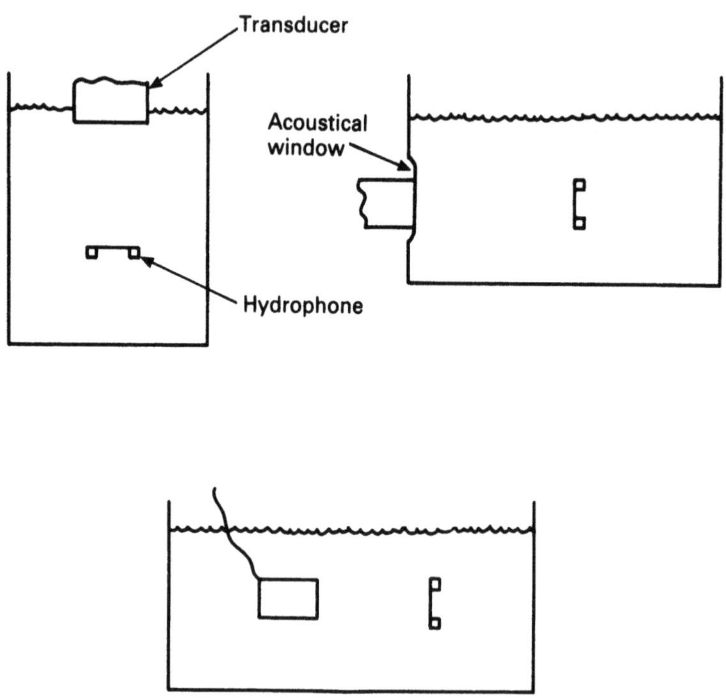

Figure 9.1. Schematic diagrams of three mounting arrangements for test-tanks.

The second method utilises a thin acoustically-transparent membrane in the side of a test-tank and uses normal coupling gel to couple the ultrasound acoustically into the tank. This method has the advantage that it does not require extra care over water penetration, however it is necessary to take account of the acoustical transfer loss which may take place across the membrane (see 9.3.9). The third method is to immerse the transducer fully in the test-tank. To do this either the transducer is sealed against water ingress or it is put in a plastic bag – a biopsy bag is often ideal for this application. Again, coupling gel is used between the transducer output face and the bag to ensure minimum acoustic transmission loss. Once immersed, the orientation can be varied according to the requirements but usually the ultrasonic field is horizontal. The NPL Beam Plotting Facility [2] utilises this type of arrangement.

Some ultrasonic diagnostic systems use a large water-bath with the transducer at the bottom or side, as do some lithotripters. To make measurements on these systems it is necessary to mount the hydrophone on a positioning system within the water tank. This may require special mounting arrangements to be devised although many of the concepts and requirements mentioned here apply in a similar manner.

9.2.2 Mechanical Clamping of the Transducer

In setting up to make acoustic output measurements of transducers it is important to ensure that the transducer is held rigid, yet without being damaged and in a sufficiently versatile manner to allow movement for alignment purposes. This is especially true for the real-time scanner transducers as they are relatively delicate and valuable. The advice given on clamping arrangements in 7.1 applies here.

Once clamped, it is useful to be able to tilt the clamp/transducer assembly for alignment purposes. In general, it is also advisable to arrange the beam-alignment axis of the transducer (the direction of the ultrasound) to be parallel to the z axis of the hydrophone positioning system. It is not possible to use an acoustic signal as the basis for setting up a real-time scanner transducer until a signal has been detected. Consequently, real-time systems are initially aligned mechanically so that the z axis of the hydrophone positioning system is parallel to the axis of symmetry of the scan plane. Often the orientation of the scan plane is obvious or is known in relation to the external body of the transducer which makes this process easy. Should this not be the case, it is convenient to operate the scanner and observe the displayed image of the test-tank. An image of the hydrophone can be of great assistance in this process, especially for sector scanners and other probes whose scan plane orientation may not always be obvious from its external appearance. After the scan plane has been identified, the transducer can be aligned so that the z-axis of the hydrophone positioning system is parallel to the axis of symmetry of the scan plan. The orientation of the transducer is then adjusted so that the scan plane is parallel to the z–x or z–y plane of the hydrophone positioning system.

After this mechanical alignment, the hydrophone should, when moved, maintain approximate alignment with the central scan line of the transducer. Once the acoustic signal has been detected refinements can be made to this alignment to improve the colinear relationship between the acoustic axis and the hydrophone positioning system z-axis.

9.2.3 Bubbles

It is easy to trap bubbles of air between a transducer and a membrane or to find bubbles on the horizontal surface of a transducer immersed with its face pointing vertically downwards. Both at the initial setting up stage and periodically during measurements, it is essential to check that no bubbles are present or appear at a later stage.

9.2.4 Measurement Methods Without an Electrical Trigger

If a measurement system consists of a test-tank and measuring hydrophone as described in Chapter 7 then the signal detected by the hydrophone will consist of a series of pulses which will either be constant in magnitude if the scanner is operating in M-mode or will vary cyclicly if the scanner is operating in B-mode (see 8.4). In the former case, it is simple to trigger the measurement system from the received signal in order to obtain measurements. Clearly, all acoustic signals are identical so it does not matter which one is used for the measurement. However, care must be exercised to ensure the signal is acoustic rather than electrical noise (pick-up). Unlike the situation using an electrical trigger (see 9.2.5), altering the relative separation of the transducer and hydrophone does not allow differentiation between these two types of signal. Confirmation of an acoustic signal is possible by moving the hydrophone laterally (in the x or y directions), an acoustic signal should change significantly whereas an electrical signal would not. Alternatively, an acoustic reflector, say a small sheet of thin expanded polystyrene, may be placed in front of the hydrophone; any acoustic signal will then be drastically reduced.

In the case of B-mode operation, it is possible to increase the setting of the threshold level for the oscilloscope/digitiser until only the largest signal is detected. This method allows the selection of just one scan line, the one giving the largest signal. Problems arise here once the hydrophone is moved, especially laterally in the x or y directions, because the signal level will change and triggering may cease altogether or may take place for a different scan line. This can easily lead to confusion and therefore errors. Overall, making measurements without a defined trigger signal is fraught with difficulty.

If, on the other hand, the only acoustical parameter of interest is I_{spta}, an alternative method using an RF power meter has been developed by Martin [3] which allows the integration of the square of the hydrophone signal for all signals received (see 7.2.4 and 7.3.3). This process automatically takes care of the need to determine the "overlap factor" (see 9.3.6) but the method is somewhat prone to drift and suffers from low signal-to-noise ratio.

9.2.5 Electrical Trigger

In order to isolate one electrical excitation pulse of the transducer and to sample the acoustic signal it generates, it is necessary to synchronise the measurement of the hydrophone signal to the generation of the acoustic pulse. This is best achieved by triggering the hydrophone measurement system using an electrical signal derived from the transducer. Such a trigger can be obtained by one of the following methods:

1. Internal sampling of the electrical excitation of the transducer using an oscilloscope probe. Manufacturer's advice should be sought here. In general, this is the best method and can usually be done with the cooperation of the manufacturer.
2. Using an electromagnetic pick-up coil placed near the transducer. Although satisfactory in many instances, the magnitude of the pick-up signal will change depending on factors such as the number of elements being fired in phased arrays, the output power setting, the focal zone selected etc. Hence, it may not be possible to obtain a stable trigger for all operating conditions of the scanner.
3. Using an auxiliary acoustic sensor placed in the acoustic field. This can be done using a small needle-like hydrophone. Like the electromagnetic method, the problem is that the signal received will depend on the operating mode of the scanner. In addition, electrical and acoustical signals will be detected by the auxiliary sensor, thus making triggering uncertain.

Even though a series of trigger pulses, one for each acoustic firing, may be available, it is now necessary to select one particular trigger pulse during successive scan repetition periods. This can be achieved with the aid of a second electrical trigger signal from the scanner, one which occurs once per frame and allows synchronisation of measurements of the hydrophone signal to the selected acoustic pulse during each scan repetition period. Either a delayed-timebase oscilloscope or a specially constructed logic counter unit can be used. Figs. 9.2 and 9.3 illustrate the two techniques.

In Fig. 9.2 it is essential to have an oscilloscope with a delayed-timebase trigger output, the delayed timebase being capable of being triggered by the vertical input (scan line trigger signal).

The two methods used for the NPL Beam Plotting Facility [2] are the delayed-timebase (scan-line and scan-frame trigger inputs) and logic counter. However, the NPL Ultrasound Beam Calibrator [1] utilises either the electromagnetic pick-up or the internal scan-line electrical signal. In the latter case, synchronisation is achieved by knowing the number of pulses in a scan frame to allow internal counting. This method only works if the number of pulses per scan frame is constant. If the scanner has different numbers of acoustic pulses occurring in consecutive scan frames the delayed-timebase or the logic methods illustrated in Figs. 9.2 and 9.3 respectively have to be used. These only require the pulse sequence after the scan frame start pulse to be reproducible, such that the nth trigger pulse always follows the same acoustic path.

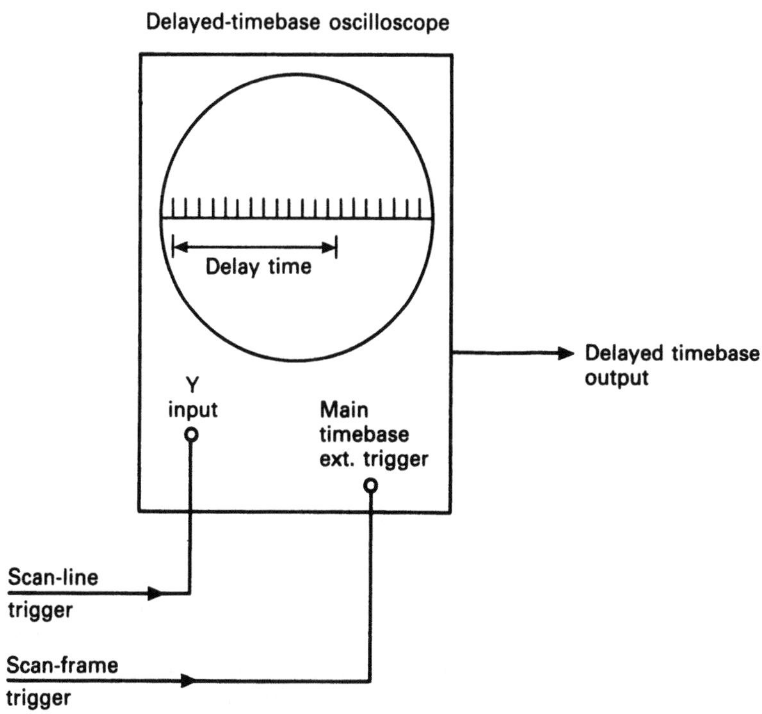

Figure 9.2. The use of a delayed-timebase oscilloscope to select a trigger signal for a chosen scan line during a scan frame for a real-time scanning system.

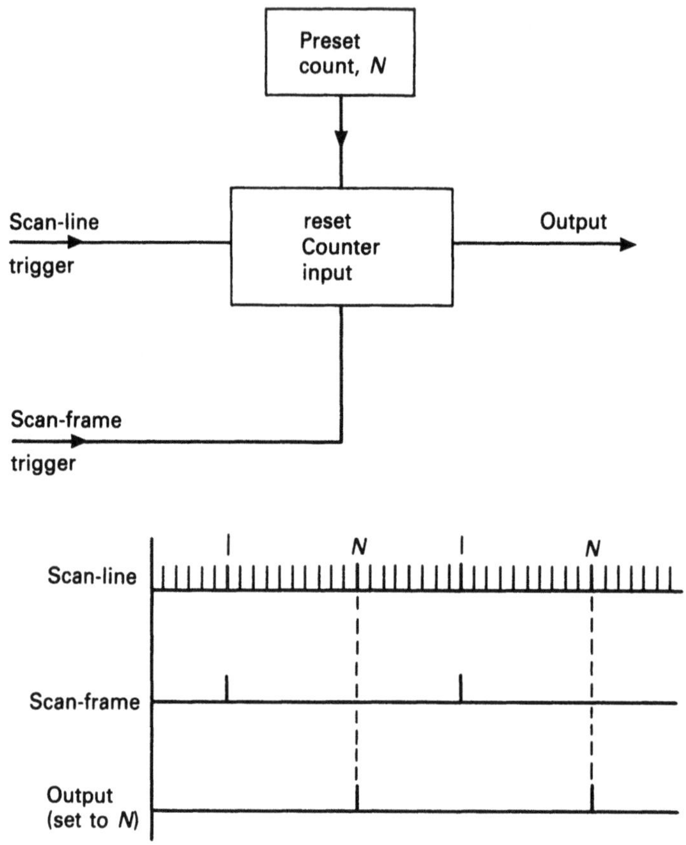

Figure 9.3. The use of a preset counter to generate a trigger signal for a chosen scan line during a scan frame for a real-time scanning system.

9.3 Making Measurements

9.3.1 Locking on to Repetitive Scan Lines

In principle, the concept of making measurements of the acoustic output of a scanning system is no different from making measurements on a fixed beam from a single-element static transducer. It is always essential to be able to separate each electrical excitation of the transducer and thereby sample the acoustic signal that it, and only it, generates. The important feature of all scanning systems is that consecutive transducer excitations will not produce the same acoustic signal at the measuring hydrophone. This is also true for multiple focal zone non-scanning systems. Again, the method of coping with this situation is the same; it is essential to be able to sample only acoustic signals that are repeating. The particular repetition rate will depend on the system and its mode of operation. Table 9.1 gives typical examples of systems, their repetition rates and the method of triggering that can be adopted.

Having isolated the repeating excitation pulse, it is used as a trigger to permit the observation of the repeating hydrophone signal. These signals can then be treated as repeating single pulses and all measurements can be made to determine acoustical parameters associated with a single pulse (see Chapter 7). For all time-average or spatial-average parameters the results are referred to as "beam" rather than "spatial". The term "beam" is being used here to refer to the fact that only one type of scan line information has been sampled; i.e. one "beam". The exception is the parameter I_{spta} which requires measurements to take account of the contributions from other scan lines (see 9.3.6).

9.3.2 Modes of Operation – Special Problems

In general, each mode of operation of a scanner can be considered as either a discrete or combined mode (see 8.4). A discrete mode is one in which only one type of information is obtained such as B-mode, M-mode or Doppler mode (D-mode). A combined mode is one in which acoustic pulses of various

Table 9.1. Examples of pulse and scan repetition rates and methods of triggering for real-time scanning systems operating in various modes

Type of ultrasonic field	Repetition rate for identical lines	Total no. of lines per scan frame	Number of discrete lines per scan frame	Method of triggering
Non-scanning single focal zone (M-mode on a linear array)	2.1 kHz	–	–	Triggered directly from the excitation pulse
Non-scanning 4 focal zone (each one fired once) (M-mode on multiple focus linear array)	4.4 kHz	–	–	Triggered every 4 excitation pulses using divide-down counter or oscilloscope in delayed trigger mode with 4 pulses displayed during the main timebase sweep
Scanning system with single focal zone firing 128 lines (B-mode on a sector scanner)	20 Hz	128	128	Triggered every 128 excitation pulses using a scan frame trigger. Alternatively, using a counter
Scanning system with 4 focal zones (each fired once) and 128 lines per frame (B-mode on a phased array scanner)	5 Hz	512	128	Triggered every 512 excitations using the same methods as previous example, above

types are produced and intermingled in order to generate two or more types of diagnostic information. Examples of combined modes would be B+M, B+D etc.

In all cases it is essential that a repeating frame sequence is identified. With the correct trigger, the sequence of signals received at the hydrophone would then repeat at the frame repetition rate. This is again done either by using a scan-frame trigger or by knowing how many acoustic pulses make up a scan frame. Some of the problems encountered here are:

1. Consecutive frames may have slightly different numbers of acoustic pulses (acoustic lines). This can sometimes be overcome by using the "start-of-frame" trigger to initiate the selection of a particular scan line by one of the two methods given in 9.2.5. This works when the sequence of pulses repeats from the start of the scan frame. Most systems have some form of synchronisation for the start of the frame even if they finish unpredictably. This means that the *n*th line would occur at the same position in the scan relative to the start but that the total number of pulses in the frame may not be the same from one frame to the next.

2. Changing the mode of operation of the scanner may change the pulsing sequence and the number of scan lines generated in a scan frame. It is necessary to re-determine the number of scan lines per frame if a counting system is being used, as for example in the case of the Beam Calibrator. The delayed time-base oscilloscope or the logic method described in 9.2.5 can be used as these are re-set for each scan frame – assuming the scan frame trigger signal is available.

3. Some systems have scan frames which are made up of two sets of scan lines, each one starting at opposite ends of the frame. For instance, a mechanical sector wobbler transducer might fire a sequence going in one direction and then reverse the sequence going back. This can be coped with by selecting every other scan frame or doubling the number of scan lines, thereby treating the to-and-fro scan as one frame. To select every other scan frame using the delayed oscilloscope method (Fig. 9.2), the main timebase sweep rate is set so that the second scan frame trigger occurs before the end of the sweep.

9.3.3 Scan Format

Ultrasonic scanning systems can generate a wide range of scan formats from simple rectilinear and sector to combined rectilinear (at different angles). Some of the simplest and commonest examples are shown in Fig. 9.4. As all acoustical measurements made using a hydrophone need to be made with the hydrophone oriented for maximum signal (they are not omni-directional sensors, see Chapter 4), it is essential to know what format of scan frame is anticipated. Knowing the format, it is then possible to decide how to position the transducer or hydrophone in order to measure any of the scan lines (see also 9.2.2). In practice, it is usual to measure the output of a typical scan line generated in the centre of the scan and to check that other scan lines give similar levels (see 9.3.4).

For measurements on a rectilinear scan format, the hydrophone (or transducer) is best moved using a linear coordinate positioning system (*x–y*

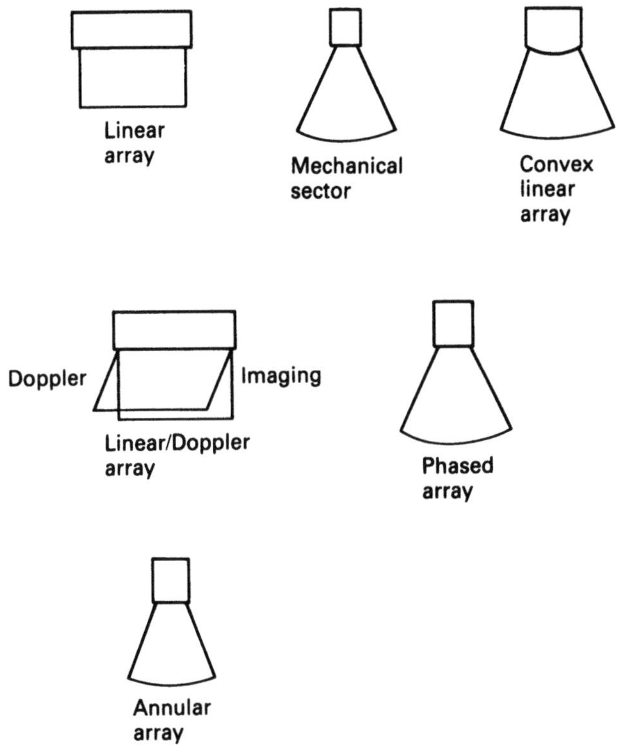

Figure 9.4. Typical scan formats produced by real-time scanners.

translation). The hydrophone rotation would not have to be changed when examining the output of different scan lines. However, for any one of the systems which generate radial scan lines the need for hydrophone rotation would depend on the orientation of the particular scan line being measured. To simplify measurements it may be better to mount the transducer so that it can be rotated about an axis corresponding to the centre of the radial pattern. This allows the hydrophone to be moved using the linear translation whilst maintaining the translation axes in line with the acoustic axis. This is illustrated in Fig. 9.5. In practice, it is usually very difficult to set up a versatile system to achieve these objectives and at the same time cope with the wide range of transducer geometries and scan formats encountered. If scan lines other than the central ones are to be measured, it is often easier to re-position the transducer in its clamping arrangement.

9.3.4 Variation Between Scan Lines

Modern scanners tend to show variations in the acoustic output of most scan lines of less than ±5% in peak pressure. However, for linear-phased arrays which excite a small group of elements and move this group sequentially along the array, the output level at the ends of the array can be significantly lower;

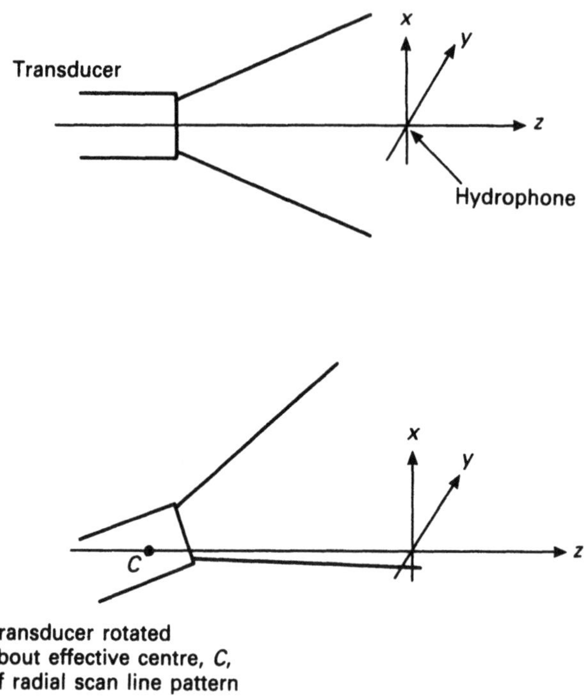

Figure 9.5. Illustration of an ideal mounting arrangement for a sector or phased array transducer in which it can be rotated about the effective centre of the scan format.

it is not unusual to find 50% lower pressure levels for the scan lines at the ends of the array.

For multiple-focus systems there will be large differences between the output levels measured at any one point in the ultrasonic field from the different focal zone firings. Each focal zone will have peak pressures and intensities at different distances from the face of the transducer.

9.3.5 Multiple Focal Zones and Multiple Firing

In phased, linear-phased or annular-phased array transducers, it is now common to find the capability to vary the phase between the electrical excitation signals applied to the elements. This provides focusing at a range of depths, commonly four focal depth settings are available either singly or in combination. Each manufacturer may produce variations.

To cope with multiple focal zones, it is again necessary to be able to select each one separately so that each can effectively be treated as a discrete mode. In fact, many scanners allow the selection of each focal zone separately. Measurements can then be undertaken on each focal zone as though it were a discrete mode. Combined focal zones can also be treated in the same way providing the method of triggering allows ready selection of the different focal zone trigger signals. When determining the value of I_{spta} it is necessary to take account of the contributions from all focal zones as described in 9.3.6. It is essential to check that the acoustical parameters measured for a particular focal zone operated separately are the same as those for the same focal zone when the system is operated in multiple focal zone mode. It is often the case that they are not.

9.3.6 Overlap Factor in Determining I_{spta}

Once the measurements on a particular scan line have been completed, it is important to take account of the contribution to the temporal-average intensity from other scan lines. This is because the beams from adjacent scan lines overlap each other. Figure 9.6 shows a schematic representation of this problem.

The true spatial-peak temporal-average intensity, I_{spta}, is determined by:

$$\sum_{n=1}^{N} I_{ta}(n)$$

where $I_{ta}(n)$ is the temporal-average intensity for the nth scan line and N is the number of scan lines per frame. If the central line measured has index i and has a temporal-average intensity I_{bpta} (beam-peak temporal-average) then the summation reduces to:

$$I_{bpta} + \sum_{\substack{n=1 \\ n \neq i}}^{N} I_{ta}(n)$$

in which the summation is called the "overlap factor".

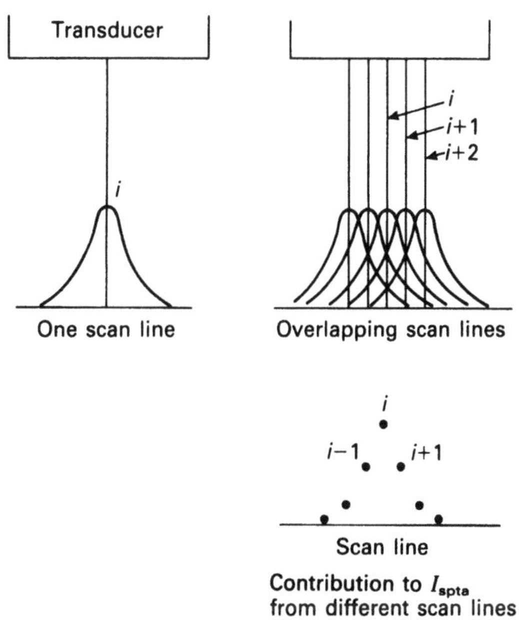

Figure 9.6. Schematic diagram showing the contributions to I_{spta} from overlapping scan lines.

To measure the "overlap factor", the hydrophone is first placed at the point of maximum temporal-average intensity for the ith line (equal to I_{bpta}). The trigger is then changed by one line and the temporal-average intensity $I_{ta}(n)$ is determined. This process is repeated systematically and all contributions from other lines are determined. The magnitude of the sum of these contributions, the "overlap factor", can vary significantly between different systems. In situations where measurements are made at the focus of a tightly focused beam, the "overlap factor" can be almost negligible. However, if the hydrophone is moved to a position where the beam is less well focused the overlap inevitably gets larger. The overlap factor can be as high as ten. This means that the spatial-peak temporal-average intensity could be underestimated by a factor of ten if this overlap is not taken into account.

Overlap also has other consequences. It is normally assumed that the location in the field which gives the maximum pulse-pressure-squared integral is the same point as that for I_{spta}. Whilst this is the case for non-scanning single focal zone systems this is not necessarily the case for fields generated by the more complex systems. As it is necessary to locate the point of I_{spta}, the overlap from neighbouring scan lines or alternatively the contribution from other focal zones must always be determined throughout the ultrasonic field. It is often found that the contributions from other focal zones can affect the final position and value of I_{spta}. Fig. 9.7 shows an axial plot of the temporal-average intensity from a system with four focal zones. Taken singly, focal zone 4 shows the maximum at a distance of 9.9 cm from the transducer face. However, after the contribution from the other three focal zones is taken into account, the maximum moves to 9.5 cm and could easily have occurred in the range 7.0 to 8.0 cm.

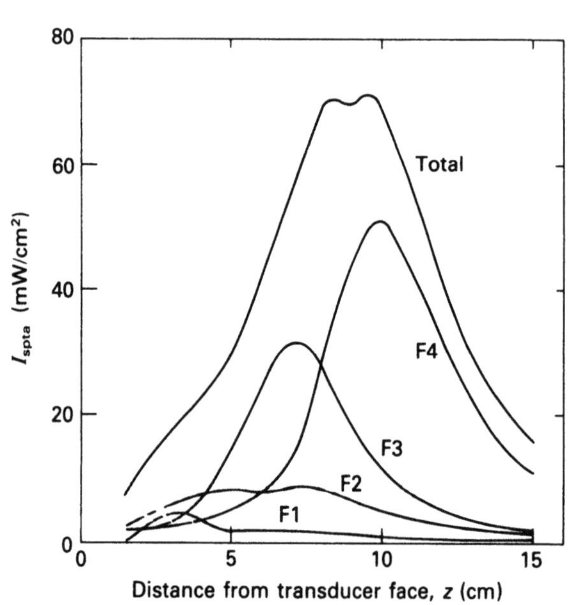

Figure 9.7. Axial variation of I_{spta} for four focal zones taken separately and in combination.

If an RF power meter is used to detect the hydrophone signal [3], then the overlap factor will automatically be taken into account as the system integrates the square of the hydrophone signal over all time. Following on from Eq. (7.9), I_{spta} will be given by:

$$I_{\text{spta}} = \frac{P_{\text{el}}R}{M_1^2\rho c}$$

where the symbols have the same meaning as in Chapter 7.

In 6.12, the concept of *in-situ* parameters was considered. It is interesting to note the potential dramatic effect that the FDA *in-situ* correction can have on a multiple focal zone system. Fig. 9.8 shows the axial plot corresponding to Fig. 9.7 after applying the *in-situ* correction. The overall maximum now occurs at 5.8 cm, very much closer to the transducer than the water maximum.

9.3.7 Asymmetrical Beam Shapes

In determining spatial-average quantities and total power by integrating the intensity over the beam (see Chapters 6 and 7) it may be appropriate or convenient to assume cylindrical symmetry of the beam in order to allow diametrical scans to be used. However, a two-dimensional raster scan should ideally be undertaken to determine the beam-area and the integrals referred to in Chapters 6 and 7. Unfortunately, such raster scans are extremely time-consuming to perform and often unnecessary. Diametrical scans are often adequate especially for symmetrical beams. Even if the beam is asymmetrical,

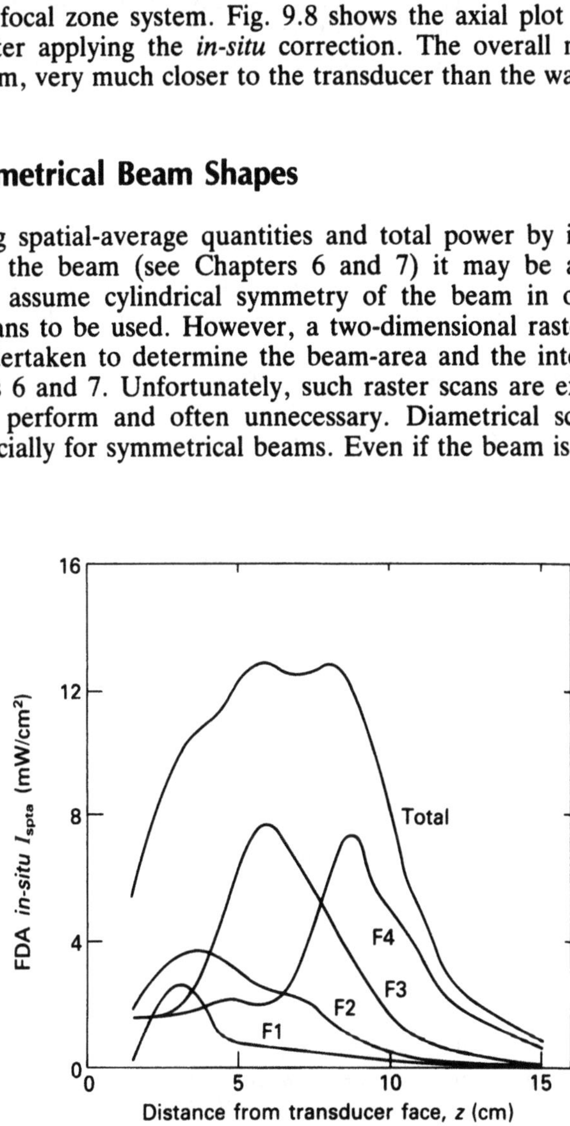

Figure 9.8 Axial variation of FDA *in-situ* levels of I_{spta} for four focal zones taken separately and in combination. To be compared with Fig. 9.7.

diametrical scans in different directions can be undertaken and an average value determined for whichever parameter is of interest.

For transducers without cylindrical symmetry, the degree of asymmetry will change significantly with the distance of the measurement plane from the transducer. It is possible for the ratio of the beam widths in two orthogonal directions to change by a factor of five over distances of 20% of the focal length. This often occurs for dynamically-focused systems where electronic focusing takes place in the scan plane and mechanical focusing is used perpendicular to the scan plane. This gives rise to two focal distances (defined here as the minimum beam-width), one for each of the two directions. Hence, it is not possible to assume a constant degree of asymmetry. For more information on the determination of beam shape information, raster scans and diametrical scans, the reader is referred to the International Electrotechnical Commission Draft Standard [4] and the US Food and Drug Administration Reporting Requirements [5].

9.3.8 Total Power

Total power can be determined from the hydrophone measurements by integrating the square of the hydrophone signal or by direct measurement using a radiation force balance (see Chapter 5).

If the hydrophone measurements are to be used, then the beam is scanned in a raster or diametrical pattern as mentioned in Chapter 6. The integral of the intensity (pressure squared) over the beam is then determined. Unless the beam has considerably extended wings, the main contribution to the integral occurs from the peak to the 10% pressure level (1% intensity). The power in the "beam" is then determined by knowing the repetition rate of the scan line (see Chapter 6). To determine the total power of a simple array or sector, it is necessary to multiply the beam power by the number of nominally similar beams, N. Thus:

$$\textit{Array power} = N \times \textit{Beam power}$$

In theory, a correction for the usual drop-off at the end of a linear phased array (see 9.3.4) should be made but the error is usually only a few per cent, significantly less than the overall accuracy of the measurements (see 9.4).

More complex modes such as multiple focal zones or combined-operating modes can all be treated by separating the series of acoustic pulses into their respective types. The total power is then determined by adding the contributions from each type of pulse.

9.3.9 Corrections

There are a number of potential sources of systematic bias in the acoustic output measurement processes dealt with in this chapter and a full appraisal of these can be found in Preston et al. [2]. For measurements of real-time systems, spatial averaging and limited bandwidth will be the most important.

Spatial averaging will not be important if a hydrophone has been chosen such that, for the measurements being made, the effective diameter of the active element of the hydrophone is less than 50% of the measured beam width. Bandwidth corrections will not be important unless the waveform is distorted by the effects of nonlinear propagation. If both these effects are important then the procedures and guidance given in Chapter 7 should be applied.

Apart from the possibility of corrections which are appropriate to the particular measurement system (say errors in amplifier gains etc), other corrections are usually negligible with the exception of the contributions of noise to the integration of the square of the hydrophone signal over the beam profile when determining total power [4].

One potential source of error is in the use of a biopsy bag or acoustic window between the transducer and measuring hydrophone. If such a membrane has been used then the transmission loss through the bag must be taken into account. This is best accomplished by an arrangement which allows the membrane to be interposed between the transducer and the measuring hydrophone so that signals with and without the membrane can be determined directly. A temporary vertical arrangement as shown in Fig. 9.1 is best suited to this type of measurement. Alternatively the transmission loss of the membrane may be determined independently using other transducers which can be immersed in water. A third option is to introduce a second bag or membrane of the same type in the acoustic path so that the actual loss can be determined for the transducer under test. This has the advantage that subtle bandwidth and other effects peculiar to the particular transducer under test will be automatically taken into account. Experience has shown that the preferred method is to determine the actual loss using either the first or the third method.

Table 9.2. Typical overall uncertainties at a confidence level of 95% for acoustic output measurements of a linear array or sector scanner using the NPL Beam Plotting Facility and the NPL Ultrasound Beam Calibrator respectively

Acoustical parameter	Uncertainties (%)	
	Beam Plotting	Beam Calibrator
p_+	13	22
p_-	8	13
I_{sppa}	17	36
I_{spta}	23	30
Beam width	5	22
Pulse duration	3	8
Total power	15	58[a]

[a]Unusually large beam asymmetry – for a symmetrical beam the uncertainty would be 30–35%.

9.4 Measurement Uncertainties

Using reasonable care and a suitably chosen hydrophone and measurement system, it should be possible to undertake measurements with accuracies in the range 15% to 30%, depending on the particular acoustical parameter. To give some idea of uncertainties, Table 9.2 gives typical overall uncertainties at a confidence level of 95% for measurements using the NPL Beam Plotting Facility [2] and the NPL Ultrasound Beam Calibrator [1] (see Chapter 7).

9.5 References

1. Preston RC. The NPL ultrasound beam calibrator, IEEE Trans Ultrason Ferroelec Freq Contr 1988; UFFC-35: 122–139
2. Preston RC, Bacon DR, Smith RA. Calibration of medical ultrasonic equipment – procedures and accuracy assessment. IEEE Trans Ultrason Ferroelec Freq Contr 1988; UFFC-35: 110–121
3. Martin K. Measurement of acoustic output parameters from medical ultrasound devices with an RF power meter system. IEEE Trans Ultrason Ferroelec Freq Contr 1988; UFFC-35: 140–145
4. IEC. The absolute calibration of hydrophones using the planar scanning technique in the frequency range 0.5 MHz to 15 MHz. To be published as an IEC Standard, Document 87(CO)4, International Electrotechnical Commission, Geneva, Switzerland, 1988
5. FDA. Guide for measuring and reporting acoustic output of diagnostic ultrasound medical devices. Document 510(k), US Department of Health and Human Services, Food and Drug Administration, HFZ-132, Rockville, MD 20857, USA, 1985

Glossary

Definitions Relevant to Medical Ultrasonic Field Measurements

The definitions below are based on those given in a number of documents. The majority come from two documents prepared for the International Electrotechnical Commission, Technical Committee 87: Ultrasonics. The first is to be published as a standard entitled "Measurement and characterisation of ultrasonic fields using hydrophones in the frequency range 0.5 MHz to 15 MHz" [IEC document 87(CO)6, 1988]. The second is a Draft International Standard entitled "Requirements for the declaration of the acoustic output of medical diagnostic ultrasonic equipment" [Document 87(CO)11, 1990]. Other definitions come from the US Food and Drug Administration 510(k) Reporting Requirements.

Acoustic Pulse Waveform The temporal waveform of the instantaneous acoustic pressure at a specified position in an acoustic field and displayed over a period sufficiently long to include all significant acoustic information in a single pulse or tone-burst, or one or more cycles in a continuous wave.

Acoustic Repetition Period Equal to the pulse repetition period for non-automatic scanning systems and the scan repetition period for automatic scanning systems. Equal to the time interval between consecutive cycles for continuous-wave systems.

Unit: second, s

Acoustic Working Frequency (AWF) The frequency of an acoustic signal based on the observation of the output of a hydrophone placed in an acoustic field at the position corresponding to the spatial-peak temporal-peak acoustic pressure. The signal is analysed either using the zero-crossing frequency technique or using a spectrum analysis method.

Symbol: f_{awf}
Unit: hertz, Hz

The following acoustic working frequencies are defined.

For the analysis based on the acoustic pulse waveform:

Zero-Crossing Acoustic Working Frequency. The inverse of the time interval between the extreme zero-crossing points of the pulse-peak cycle. For cw systems, the inverse of the time interval between three consecutive zero-crossing points in the acoustic pulse waveform.

For the spectral analysis method the acoustic pressure versus frequency is displayed and there are three different methods of deriving the acoustic working frequency which are:

Arithmetic-Mean Acoustic Working Frequency (Centre Frequency). The arithmetic mean of the frequencies f_1 and f_2 at which the amplitude of the spectrum of the acoustic pressure first becomes 3 dB lower than the peak amplitude.

Geometric-Mean Acoustic Working Frequency. The geometrical mean of the frequencies f_1 and f_2.

Modal Acoustic Working Frequency. The frequency corresponding to the maximum amplitude in the acoustic pressure spectrum.

Note: Arithmetic-mean acoustic working frequency is equivalent to center frequency in the US FDA 510(k) Reporting Requirements.

Bandwidth Difference in the frequencies f_1 and f_2 at which the amplitude of the spectrum of the acoustic pressure first becomes 3 dB lower than the peak amplitude.

Beam-Alignment Axis This parameter is used for alignment purposes only, and is a straight line joining two points of spatial-peak temporal-peak acoustic pressure on two hemispherical surfaces whose centres are at the approximate geometrical centre of an ultrasonic transducer or ultrasonic transducer element group. One hemisphere has a radius of curvature of approximately $A_g/\pi\lambda$, where A_g is the geometrical area of the ultrasonic transducer or ultrasonic transducer element group and λ is the wavelength of the ultrasound corresponding to the nominal frequency. The second hemisphere has a radius of curvature either $2A_g/\pi\lambda$, or $A_g/3\pi\lambda$ whichever is the more appropriate. For the purposes of alignment, this line may be projected to the face of the ultrasonic transducer or ultrasonic transducer element group.

For most practical applications, two plane surfaces perpendicular to the direction of propagation of the ultrasound are used. In cases where a unique peak is not located on a hemispherical surface, another hemispherical surface is chosen with a different radius of curvature yielding a unique peak.

Beam Area The beam-area is the area in a specified surface consisting of all points at which the pulse-pressure-squared integral is greater than a specified fraction of the maximum value of the pulse-pressure-squared integral in that surface. If the position of the surface is not specified, it is the surface passing through the point of spatial-peak temporal-peak acoustic pressure in the whole acoustic field. The specified levels are 0.25 and 0.01 for the −6 dB and −20 dB beam-areas respectively.

Note 1: The beam-area may be composed of several sections.
Note 2: The specified surface is usually a plane perpendicular to the beam-alignment axis but can be a cylinder for ultrasonic transducers with cylindrical active elements or a sphere for ultrasonic transducers with spherical active elements.

Symbol: A_b
Unit: metre squared, m^2

Beam-Average Pulse Acoustic Pressure The pulse acoustic pressure from one ultrasonic transducer or ultrasonic transducer element group averaged over the −6 dB beam-area in a specified plane or in a plane containing the spatial-peak temporal-peak acoustic pressure for that particular ultrasonic transducer or ultrasonic transducer element group.

Note: −6 dB beam-area is commonly used. However, other beam-areas may be used (see beam-area definition).

Symbol: p_{bap}
Unit: pascal, Pa

Beam-Average Pulse-Average Intensity The pulse-average intensity from one ultrasonic transducer or ultrasonic transducer element group averaged over the −6 dB beam-area in a specified plane or in a plane containing the spatial-peak temporal-peak acoustic pressure for that particular ultrasonic transducer or ultrasonic transducer element group.

Note: −6 dB beam-area is commonly used. However, other beam-areas may be used (see beam-area definition).

Symbol: I_{bapa}
Unit: watt per metre squared, W/m^2

Beam-Average RMS Acoustic Pressure The rms acoustic pressure from one ultrasonic transducer or ultrasonic transducer element group averaged over the −6 dB beam-area in a specified plane or in a plane containing the spatial-peak temporal-peak acoustic pressure for that particular ultrasonic transducer or ultrasonic transducer element group.

Note: −6 dB beam-area is commonly used. However, other beam-areas may be used (see beam-area definition).

Symbol: p_{bar}
Unit: pascal, Pa

Beam-Average Temporal-Average Intensity The temporal-average intensity from one ultrasonic transducer or ultrasonic transducer element group averaged over the −6 dB beam-area in a specifed plane or in a plane containing the spatial-peak temporal-peak acoustic pressure for that particular ultrasonic transducer or ultrasonic transducer element group.

Note: −6 dB beam-area is commonly used. However, other beam-areas may be used (see beam-area definition).

Symbol: I_{bata}
Unit: watt per metre squared, W/m^2

Central Scan Line For automatic scanning systems, the ultrasonic scan line closest to the symmetry axis of the scan plane.

Combined-Operating Mode Mode of operation of a system which combines more than one discrete-operating mode.

Note: Examples of combined-operating modes are real-time B-mode combined with M-mode (B+M), real-time B-mode combined with pulsed Doppler (B+D), real-time B-mode combined with M-mode and pulsed Doppler (B+M+D), real-time B-mode combined with real-time flow mapping Doppler (B+rD) i.e. flow-mapping in which different types of acoustic pulses are used to generate the Doppler information and the imaging information.

Discrete-Operating Mode Mode of operation of medical diagnostic ultrasonic equipment in which the purpose of the excitation of the ultrasonic transducer or ultrasonic transducer element group is to generate only one type of clinical information.

Note: Examples of discrete-operating modes are A-mode (A), M-mode (M), static B-mode (sB), real-time B-mode (B), cw Doppler (cwD), pulsed Doppler (D), static flow-mapping (sD) and real-time flow-mapping (rD) using only one type of acoustic pulse.

Effective Area of an Ultrasonic Transducer The area of a perfect piston-like ultrasonic transducer which has a predicted axial acoustic pressure distribution approximately equivalent to the observed axial acoustic pressure distribution over a limited axial distance.

Symbol: A_1
Unit: metre squared, m^2

Effective Radius of a Hydrophone Active Element The radius of a stiff disc receiver hydrophone which has a predicted directional response function with an angular width equal to the observed angular width. The angular width is determined at a specified level below the peak of the directional response function. For the specified levels of 3 dB and 6 dB, the radii are denoted by a_3 and a_6 respectively.

Symbols: a, a_3, a_6
Unit: metre, m

End-of-Cable Loaded Sensitivity of a Hydrophone The ratio of the voltage at the end of any integral cable or connector of a hydrophone, when connected to a specified electrical input impedance, to the instantaneous acoustic pressure in the undisturbed free field of a plane wave in the position of the acoustic centre of the hydrophone if the hydrophone were removed.

Symbol: M_l
Unit: volt per pascal, V/Pa

End-of-Cable Open-Circuit Sensitivity of a Hydrophone The ratio of the open-circuit voltage at the end of any integral cable or connector of a hydrophone to the instantaneous acoustic pressure in the undisturbed free field of a plane wave in the position of the acoustic centre of the hydrophone if the hydrophone were removed.

Symbol: M_c
Unit: volt per pascal, V/Pa

Far Field The acoustic (sound) field at distances from an ultrasonic transducer where the values of the instantaneous acoustic pressure and particle velocity are substantially in phase.

For most purposes, although strictly for planar ultrasonic transducers, the far field is at a distance greater than $A_1/\pi\lambda$, where A_1 is the effective area of the ultrasonic transducer or ultrasonic transducer element group and λ is the wavelength of the ultrasound corresponding to the acoustic working frequency.

Geometrical Area of an Ultrasonic Transducer The area defined by the ultrasonic transducer dimensions or ultrasonic transducer element group dimensions.

Symbol: A_g
Unit: metre squared, m^2

Geometrical Radius of a Hydrophone Active Element The radius defined by the dimensions of the active element of a hydrophone.

Symbol: a_g
Unit: metre, m

Hydrophone A transducer that produces electrical signals in response to waterborne acoustic signals.

Inclusive Mode Combined-operating mode having acoustic output levels (p_- and I_{spta}) less than those corresponding to a specified discrete-operating mode.

Instantaneous Acoustic (or Sound) Pressure The value of the pressure amplitude minus the ambient pressure at a particular instant in time and at a particular point in an acoustic field.

Symbol: p
Unit: pascal, Pa

Instantaneous Intensity The acoustic energy transmitted per unit time in the direction of acoustic wave propagation per unit area normal to this direction at a particular instant in time and at a particular point in an acoustic field.

It is not possible to measure the true instantaneous intensity. However, for historical reasons, it is preferable to derive certain intensity parameters from hydrophone measurements. Therefore, for the measurement purposes, and if it is reasonable to assume far field conditions, the instantaneous intensity, I is expressed as

$$I = \frac{p^2}{\rho c},$$

where p is the instantaneous acoustic pressure, ρ is the density of the medium and c is the speed of sound in the medium.

Symbol: I
Unit: watt per metre squared, W/m^2

Maximum Intensity The instantaneous intensity averaged over the half cycle containing the temporal-peak acoustic pressure.

Symbol: I_m
Unit: watt per metre squared, W/m^2

Mean-Peak-Cycle Acoustic Pressure The arithmetic mean of the maximum positive and modulus of the maximum negative instantaneous acoustic pressure during a pulse-peak cycle.

Symbol: p_m
Unit: pascal, Pa

Medical Diagnostic Ultrasonic Equipment (or System) Combination of the ultrasonic instrument console and the transducer assembly making up a complete diagnostic system.

Near Field The near field is the region of the acoustic (sound) field near an ultrasonic transducer where the relative phase of the instantaneous acoustic pressure and the particle velocity is continually changing with position in the acoustic (sound) field.

For planar transducers, this is at a distance less than $A_1/\pi\lambda$, where A_1 is the effective area of the ultrasonic transducer or ultrasonic transducer element group and λ is the wavelength of the ultrasound corresponding to the acoustic working frequency.

Nominal Frequency The ultrasonic frequency of operation of an ultrasonic transducer or ultrasonic transducer element group quoted by the designer or manufacturer.

Nonlinear Propagation Parameter An index which permits the prediction of nonlinear distortion of ultrasound for a specific ultrasonic transducer, and is given by σ_m from:

$$\sigma_m = \frac{\beta \omega l_1}{\rho c^3} p_m \frac{1}{(F_g - 1)^{\frac{1}{2}}} \ln \left((F_g - 1)^{\frac{1}{2}} + F_g^{\frac{1}{2}} \right)$$

where β is the nonlinearity parameter ($\beta = 1 + \frac{B}{2A} = 3.5$ for pure water at 20°C), ω is the angular frequency ($\omega = 2\pi f_{awf}$ where f_{awf} is the acoustic working frequency), l_1 is the distance from the face of the ultrasonic transducer to the plane containing the point of spatial-peak temporal-peak acoustic pressure and F_g is 0.69 times the ratio of the geometrical area of the ultrasonic transducer to the −6 dB beam-area. p_m is the mean-peak-cycle acoustic pressure at the point in the acoustic field corresponding to the spatial-peak temporal-peak acoustic pressure.

Symbol: σ_m

Note: The equation given above is applicable to ultrasonic fields in which $F_g > 2.1$. The specification of an index to cover $F_g \leq 2.1$ is under consideration.

Non-Scanning Mode A mode of operation of medical diagnostic ultrasonic equipment (system) that involves a sequence of ultrasonic pulses which give rise to ultrasonic scan lines that follow the same acoustic path.

Output Beam-Area Area of the ultrasonic beam derived from the output beam dimensions.

Unit: metre squared, m^2

Output Beam Dimensions The dimensions of the ultrasonic beam (−6 dB pulse beam-width) in a direction normal to the beam-alignment axis and at the transducer output face.

Output Beam Intensity Temporal-average power output divided by the output beam-area.

Symbol: I_{ob}
Unit: watt per metre squared, W/m^2

Peak-Negative (or Peak-Rarefactional) Acoustic Pressure The maximum of the modulus of the negative instantaneous acoustic pressure in an acoustic field or in a specified plane during an acoustic repetition period. Peak-negative acoustic pressure is expressed as a positive number.

Symbol: p_- (or p_r)
Unit: pascal, Pa

Peak-Positive (or Peak-Compressional) Acoustic Pressure The maximum positive instantaneous acoustic pressure in an acoustic field or in a specified plane during an acoustic repetition period.

Symbol: p_+ (or p_c)
Unit: pascal, Pa

Pulse Acoustic Pressure The square root of the ratio of the pulse-pressure-squared integral to the pulse duration at a particular point in an acoustic field.

Symbol: p_p
Unit: pascal, Pa

Pulse-Average Intensity The ratio of the pulse-intensity integral to the pulse duration at a particular point in an acoustic field.

Unit: watt per metre squared, W/m^2

Pulse Beam-Width Distance between two points, on a specified surface and in a specified direction passing through the point of the maximum pulse-pressure-squared integral (ppsi) in that surface, at which the ppsi is a specified fraction of the maximum value of the ppsi in the surface. The two points are farthest from and on opposite sides of the point of maximum ppsi. If the position of the specified surface is not given, then it passes through the point of spatial-peak temporal-peak acoustic pressure in the whole acoustic field.

The specified levels are 0.25 and 0.01 for the −6 dB and −20 dB pulse beam-widths respectively.

Note 1: For cw systems, the term pulse-pressure-squared integral (ppsi) is replaced everywhere in the above definition by mean square acoustic pressure.
Note 2: The specified surface is usually a plane perpendicular to the beam-alignment axis but can be a cylinder for ultrasonic transducers with cylindrical active elements or a sphere for ultrasonic transducers with spherical active elements.

Symbol: w_{pb6}, w_{pb20}
Unit: metre, m

Pulse Duration Equal to 1.25 times the interval between the time when the time integral of the square of the instantaneous acoustic pressure reaches 10 percent and 90 percent of its final value. The final value of the time integral of the square of the instantaneous acoustic pressure is the pulse-pressure-squared integral.

Symbol: t_d
Unit: second, s

Pulse-Intensity Integral The time integral of the instantaneous intensity at a particular point in an acoustic field integrated over the acoustic pulse waveform.

Note: For many measurement purposes, pulse-intensity integral is proportional to pulse-pressure-squared integral.

Symbol: I_{pi}
Unit: joule per metre squared, J/m^2

Pulse-Peak Cycle A single cycle of an acoustic pulse waveform between two points of zero instantaneous acoustic pressure. This single cycle is composed of the half cycle containing the temporal-peak acoustic pressure and whichever of the two adjacent half cycles contains the larger peak instantaneous acoustic pressure.

Pulse-Pressure-Squared Integral The time integral of the square of the instantaneous acoustic pressure at a particular point in an acoustic field integrated over the acoustic pulse waveform.

Symbol: p_i
Unit: pascal squared second, Pa^2s

Pulse Repetition Period The time interval between two successive pulses or tone-bursts. This applies to single element non-automatic scanning systems and automatic scanning systems.

Unit: second, s

Pulse Repetition Rate The inverse of the pulse repetition period.

Symbol: *prr*
Unit: hertz, Hz

Reference Direction For scanning modes, the direction normal to the beam-alignment axis for an ultrasonic scan line and in the scan plane. For non-scanning modes, the direction normal to the beam-alignment axis and parallel to the direction of maximum −6 dB pulse beam-width.

RMS Acoustic Pressure The root mean square of the instantaneous acoustic pressure at a particular point in an acoustic field. The mean should be taken over an integral number of acoustic repetition periods unless otherwise specified.

Symbol: p_{rms}
Unit: pascal, Pa

Scan-Area For automatic scanning systems, the area on the surface considered consisting of all points within the beam-area of any beam passing through the surface during the scan repetition period.

Symbol: A_S
Unit: metre squared, m^2

Scanning Mode A mode of operation of medical diagnostic ultrasonic equipment (system) which involves a sequence of ultrasonic pulses which give rise to ultrasonic scan lines which do not follow the same acoustic path.

Scan Plane For automatic scanning systems, a plane containing all the ultrasonic scan lines.

Scan Repetition Period The time interval between identical points on two successive frames, sectors or scans. Only applies to automatic scanning systems.

Unit: second, s

Scan Repetition Rate The inverse of the scan repetition period.

Symbol: *srr*
Unit: hertz, Hz

Spatial-Average Pulse Acoustic Pressure For non-automatic scanning systems, equal to the beam-average pulse acoustic pressure.

Note: Spatial-average pulse acoustic pressure does not apply to automatic scanning systems, beam-average pulse acoustic pressure applies only.

Symbol: p_{sap}
Unit: pascal, Pa

Spatial-Average Pulse-Average Intensity For non-automatic scanning systems, equal to the beam-average pulse-average intensity.

Note: Spatial-average pulse-average intensity does not apply to automatic scanning systems, beam-average pulse-average intensity applies only.

Symbol: I_{sapa}
Unit: watt per metre squared, W/m^2

Spatial-Average RMS Acoustic Pressure For non-automatic scanning systems, equal to the beam-average rms acoustic pressure. For automatic scanning systems, equal to the rms acoustic pressure averaged over the scan-area, where the rms acoustic pressure is taken over the scan repetition period.

Symbol: p_{sar}
Unit: pascal, Pa

Spatial-Average Temporal-Average Intensity For non-automatic scanning systems, equal to the beam-average temporal-average intensity. For automatic scanning systems, equal to the temporal-average intensity averaged over the scan-area where the temporal-average intensity is taken over the scan repetition period.

Symbol: I_{sata}
Unit: watt per metre squared, W/m^2

Spatial-Peak Pulse Acoustic Pressure The maximum value of the pulse acoustic pressure in an acoustic field or in a specified plane.

Symbol: p_{spp}
Unit: pascal, Pa

Spatial-Peak Pulse-Average Intensity The maximum value of the pulse-average intensity in an acoustic field or in a specified plane.

Symbol: I_{sppa}
Unit: watt per metre squared, W/m^2

Spatial-Peak Pulse-Intensity Integral The maximum value of the pulse-intensity integral in an acoustic field or in a specified plane.

Symbol: I_{sppi}
Unit: joule per metre squared, J/m^2

Spatial-Peak RMS Acoustic Pressure The maximum value of the rms acoustic pressure in an acoustic field or in a specified plane.

Symbol: p_{spr}
Unit: pascal, Pa

Spatial-Peak Temporal-Average Intensity The maximum value of the temporal-average intensity in an acoustic field or in a specified plane.

Symbol: I_{spta}
Unit: watt per metre squared, W/m^2

Spatial-Peak Temporal-Peak Acoustic Pressure (or Peak Sound Pressure) The larger of the peak-positive acoustic pressure or the peak-negative acoustic pressure.

Symbol: p_{sptp}
Unit: pascal, Pa

Spatial-Peak Temporal-Peak Intensity The maximum value of the temporal-peak intensity in an acoustic field or in a specified plane.

Symbol: I_{sptp}
Unit: watt per metre squared, W/m^2

Symmetry Axis of the Scan Plane An imaginary line in the scan plane on each side of which an equal number of ultrasonic scan lines occur.

Temporal-Average Intensity The time-average of the instantaneous intensity at a particular point in an acoustic field. The time-average is taken over an integral number of acoustic repetition periods, unless otherwise specified.

Symbol: I_{ta}
Unit: watt per metre squared, W/m^2

Temporal-Average Power Time-average output power radiated by an ultrasonic transducer into an approximately free field under specified conditions in a specified medium, usually water.

Symbol: W or P
Unit: watt, W

Temporal-Peak Acoustic Pressure The maximum value of the modulus of the instantaneous acoustic pressure at a particular point in an acoustic field.

Symbol: p_0
Unit: pascal, Pa

Temporal-Peak Intensity The maximum value of the instantaneous intensity at a particular point in an acoustic field.

Unit: watt per metre squared, W/m^2

Transducer Assembly Those parts of medical diagnostic ultrasonic equipment comprising the ultrasonic transducer and/or ultrasonic transducer element group together with any integral components such as acoustic lens, in-built stand-off etc. The transducer assembly is usually separable from the control unit.

Transducer Output Face External surface of a transducer assembly which is either directly in contact with the patient or is in contact with a water or fluid path to the patient.

Ultrasonic Scan Line For automatic scanning systems, the beam-alignment axis for a particular ultrasonic transducer element group, or for a single or multiple excitation of an ultrasonic transducer or ultrasonic transducer element group.

Ultrasonic Scan Line Separation For automatic scanning systems, the distance between the points of intersection of two consecutive ultrasonic scan lines of the same type and a specified plane.

Symbol: s_s
Unit: metre, m

Ultrasonic Transducer A device capable of converting electrical energy to mechanical energy within the ultrasonic frequency range and/or reciprocally of converting mechanical energy to electrical energy.

Ultrasonic Transducer Dimensions The dimensions of the surface of the element or elements of an ultrasonic transducer.

Unit: metre, m

Ultrasonic Transducer Element Group A group of elements of an ultrasonic transducer which are excited together in order to produce a single acoustic pulse.

Ultrasonic Transducer Element Group Dimensions The dimensions of the surface of the group of elements of an ultrasonic transducer element group which includes the distance between the elements, hence representing the overall dimensions.

Unit: metre, m

Ultrasound Instrument Console Electronic unit to which the transducer assembly is attached.

Subject Index

MIX
Papier aus verantwortungsvollen Quellen
Paper from responsible sources
FSC® C105338

If you have any concerns about our products,
you can contact us on
ProductSafety@springernature.com

In case Publisher is established outside the EU,
the EU authorized representative is:
Springer Nature Customer Service Center GmbH
Europaplatz 3, 69115 Heidelberg, Germany

Printed by Libri Plureos GmbH
in Hamburg, Germany